W9-DER-553

MONOGRAPHS ON
STATISTICS AND APPLIED PROBABILITY

General Editors

D.R. Cox, D.V. Hinkley, N. Reid, D.B. Rubin and B.W. Silverman

(Full details concerning this series are available from the Publishers.)

CONTENTS

Statistics for Long-Memory Processes

JAN BERAN

Department of Economics and Statistics
University of Konstanz, Germany

CHAPMAN & HALL
P An International Thomson Publishing Company

New York • Albany • Bonn • Boston • Cincinnati
• Detroit • London • Madrid • Melbourne • Mexico City
• Pacific Grove • Paris • San Francisco • Singapore
• Tokyo • Toronto • Washington

Copyright © 1994
By Chapman & Hall
A division of International Thompson Publishing Inc.
I(T)P The ITP logo is a trademark under license

Printed in the United States of America

For more information, contact:

Chapman & Hall
One Penn Plaza
New York, NY 10119

Chapman & Hall
2-6 Boundary Row
London SE1 8HN

International Thompson Publishing Europe
Berkshire House 168-173
High Holborn
London WC1V 7AA
England

International Thomson Editores
Campos Eliseos 385, Piso 7
Col. Polanco
11560 Mexico D.F. Mexico

Thomas Nelson Australia
102 Dodds Street
South Merlbourne, 3205
Victoria, Australia

International Thomson Publishing Gmbh
Königwinterer Strasse 418
53227 Born
Germany

Nelson Canada
1120 Birchmount Road
Scarborough, Ontario
Canada, M1K 5G4

International Thomson Publishing Asia
221 Henderson Road
#05-10 Henderson Building
Singapore 0315

International Thomson Publishing Japan
Hirakawacho-cho Kyowa Building, 3F
2-2-1 Hirakawacho-cho
Chiyoda-ku, Tokyo 102
Japan

1 2 3 4 5 6 7 8 9 10 XXX 01 00 99 97 96 95 94

Library of Congress Cataloging-in-Publication Data

Beran, Jan, 1959-
    Statistics for long-memory processes / Jan Beran
        P.   cm.
    Includes bibliographical references and indexes.

    1. Mathematical statistics.   2. Stochastic processes.   I. Title.
QA276.B395 1994
519.5—dc20   94-14144
                                    CIP
    ISBN 0-412-04901-5

Please send your order for this or any Chapman & Hall book to **Chapman & Hall, 29 West 35th Street, New York, NY 10001, Attn: Customer Service Department.** You may also call our Order Department at 1-212-244-3336 or fax your purchase order to 1-800-248-4724.

For a complete listing of Chapman & Hall's titles, send your requests to **Chapman & Hall, Dept. BC, One Penn Plaza, New York, NY 10119.**

Cont

vi

Statistics for Long-Memory Processes

JAN BERAN

Department of Economics and Statistics
University of Konstanz, Germany

CHAPMAN & HALL
ITP An International Thomson Publishing Company

New York • Albany • Bonn • Boston • Cincinnati
• Detroit • London • Madrid • Melbourne • Mexico City
• Pacific Grove • Paris • San Francisco • Singapore
• Tokyo • Toronto • Washington

Copyright © 1994
By Chapman & Hall
A division of International Thompson Publishing Inc.
I(T)P The ITP logo is a trademark under license

Printed in the United States of America

For more information, contact:

Chapman & Hall
One Penn Plaza
New York, NY 10119

Chapman & Hall
2-6 Boundary Row
London SE1 8HN

International Thompson Publishing Europe
Berkshire House 168-173
High Holborn
London WC1V 7AA
England

International Thomson Editores
Campos Eliseos 385, Piso 7
Col. Polanco
11560 Mexico D.F. Mexico

Thomas Nelson Australia
102 Dodds Street
South Merlbourne, 3205
Victoria, Australia

International Thomson Publishing Gmbh
Königwinterer Strasse 418
53227 Bonn
Germany

Nelson Canada
1120 Birchmount Road
Scarborough, Ontario
Canada, M1K 5G4

International Thomson Publishing Asia
221 Henderson Road
#05-10 Henderson Building
Singapore 0315

International Thomson Publishing Japan
Hirakawacho-cho Kyowa Building, 3F
2-2-1 Hirakawacho-cho
Chiyoda-ku, Tokyo 102
Japan

1 2 3 4 5 6 7 8 9 10 XXX 01 00 99 97 96 95 94

Library of Congress Cataloging-in-Publication Data

Beran, Jan, 1959-
    Statistics for long-memory processes / Jan Beran
        P.   cm.
    Includes bibliographical references and indexes.

    1. Mathematical statistics. 2. Stochastic processes. I. Title.
QA276.B395 1994
519.5—dc20   94-14144
    ISBN 0-412-04901-5   CIP

Please send your order for this or any Chapman & Hall book to **Chapman & Hall, 29 West 35th Street, New York, NY 10001, Attn: Customer Service Department.** You may also call our Order Department at 1-212-244-3336 or fax your purchase order to 1-800-248-4724.

For a complete listing of Chapman & Hall's titles, send your requests to **Chapman & Hall, Dept. BC, One Penn Plaza, New York, NY 10119.**

Contents

Preface

The phenomenon of long memory had been known long before suitable stochastic models were developed. Scientists in diverse fields of statistical applications observed empirically that correlations between observations that are far apart (in time or space) decay to zero at a slower rate than one would expect from independent data or data following classic ARMA- or Markov-type models. As a result of Mandelbrot's pioneering work, self-similar and related stationary processes with long memory were later introduced to statistics, to provide a sound mathematical basis for statistical inference. Since then, long memory (or long-range dependence) has become a rapidly developing subject. Because of the diversity of applications, the literature on the topic is broadly scattered in a large number of journals, including those in fields such as agronomy, astronomy, chemistry, economics, engineering, environmental sciences, geosciences, hydrology, mathematics, physics, and statistics. This book grew out of an attempt to give a concise summary of the statistical methodology for data with long-range dependence, in a way that would be useful for those who need to analyze such data, as well as for those who would like to know more about the mathematical foundations of the area. To make the book readable, a selection of important topics was necessary. An extensive bibliography given at the end of the book should provide help for studying topics not covered here or not covered in detail.

For readers familiar with the topic, the chapters can be read in arbitrary sequence. For those who are new to the area, the first three chapters give the basic knowledge required to read the rest of the book. Some elementary knowledge on time series analysis is helpful. The first chapter gives a simple introduction to the topic of long-range dependence, discusses data examples, and gives a short historic overview. In the second chapter, stochastic processes for modelling long-range dependence are introduced. Chapter 3 summarizes some limit theorems needed in the chapters that follow. Estimation of the long-memory parameter (and other parameters characterizing the dependence) is discussed in Chapter 4 (simple heuristic methods), Chapters 5 and 6 (maximum likelihood estimation) and Chapter 7 (robust estimation). Location and scale estimation, as well as forecasting are discussed in Chapter 8. Chapter 9 covers regression and analysis of variance. Goodness-of-fit tests are treated in Chapter 10. Miscellaneous topics, including infinite variance processes, GARMA processes and simulation are discussed briefly in Chapter 11. Listings of a few Splus-programs and data sets are given in Chapter 12. The programs are simple straightforward implementations of selected methods discussed in the book. Instead of optimizing the programs in terms of fastest performance, the emphasis was put on simplicity and readability of the code.

I would like to thank the editor, John Kimmel, for his continuous support and encouragement. I am grateful to the following colleagues for providing me with data sets and detailed information regarding their data: H. Heeke and H. Hundt (Siemens, Munich, Germany), W. Willinger, W. Leland, D. Wilson, A. Tabatabai (Bellcore, Morristown, USA), C. Croarkin, C. Hagwood, M. Pollak and R.N. Varner (NIST, Gaithersburg, USA), H.P. Graf (CIBA Geigy, Basel, Switzerland), K.R. Briffa and P.D. Jones (Climate Research Unit, University of East Anglia, Norwich, UK) and R.L. Smith (University of North Carolina, Chapel Hill, USA). I would also like to thank Raj Bhansali, Stamatis Cambanis, Andrey Feuerverger, Chris Field, Sucharita Ghosh, Hanspeter Graf, Frank Hampel, Cliff Hurvich, Hansruedi Künsch, Doug Martin, Emanuel Parzen, Bob Shumway, Richard Smith, Werner Stahel, Paul Switzer, Murad Taqqu, Norma Terrin, Yoshihiro Yajima, Victor Yohai, Ruben Zamar and Scott Zeger for many interesting discussions on time series and long memory in particular. Also, I would like to thank three anonymous referees and my colleagues at the University of Zurich, Ueli Helfenstein and Alois Tschopp, for their valuable remarks on the first version of the manuscript. Finally, I would like to thank my family, in particular my mother and my wife, for their support and my daughter for her cheerfulness, without which this book could not have been written.

Jan Beran,
Zurich, February 1994

CHAPTER 1

Introduction

1.1 An elementary result in statistics

One of the main results taught in an introductory course in statistics is: *The variance of the sample mean is equal to the variance of one observation divided by the sample size.* In other words, if $X_1, ..., X_n$ are observations with common mean $\mu = E(X_i)$ and variance $\sigma^2 = \text{var}(X_i) = E[(X_i - \mu)^2]$, then the variance of $\bar{X} = n^{-1} \sum_{i=1}^{n} X_i$ is equal to

$$\text{var}(\bar{X}) = \sigma^2 n^{-1}. \tag{1.1}$$

A second elementary result one learns is: *The population mean is estimated by $\bar{X}$, and for large enough samples the $(1 - \alpha)$-confidence interval for μ is given by*

$$\bar{X} \pm z_{\frac{\alpha}{2}} \, \sigma \, n^{-\frac{1}{2}} \tag{1.2}$$

if σ^2 is known and

$$\bar{X} \pm z_{\frac{\alpha}{2}} \, s \, n^{-\frac{1}{2}} \tag{1.3}$$

if σ^2 has to be estimated. Here $s^2 = (n - 1)^{-1} \sum_{i=1}^{n} (X_i - \bar{X})^2$ is the sample variance and $z_{\frac{\alpha}{2}}$ is the upper $(1 - \frac{\alpha}{2})$ quantile of the standard normal distribution.

Frequently, the assumptions that lead to $(1.1), (1.2)$, and (1.3) are mentioned only briefly. The formulas are very simple and can even be calculated by hand. It is therefore tempting to use them in an automatic way, without checking the assumptions under which they were derived. How reliable are these formulas really in practical applications? In particular, is (1.1) always exact or at least a good approximation to the actual variance of $\bar{X}$? Is the probability that (1.2) and (1.3) respectively contain the true value μ always equal to or at least approximately equal to $1 - \alpha$?

In order to answer these questions one needs to analyze some typical data sets carefully under this aspect. Before doing that (in

Section 1.4), it is useful to think about the conditions that lead to
(1.1), (1.2) and (1.3), and about why these rules might or might
not be good approximations.

Suppose that $X_1, X_2, ..., X_n$ are observations sampled randomly
from the same population at time points $i = 1, 2, ..., n$. Thus,
$X_1, ..., X_n$ are random variables with the same (marginal) distribution F. The index i does not necessarily denote time. More generally, i can denote any other natural ordering, such as for example,
the position on a line in a plane.

Consider first equation (1.1). A simple set of conditions under
which (1.1) is true can be given as follows:

1. The population mean $\mu = E(X_i)$ exists and is finite.
2. The population variance $\sigma^2 = \text{var}(X_i)$ exists and is finite.
3. $X_1, ..., X_n$ are uncorrelated, i.e.,

$$\rho(i, j) = 0 \quad \text{for } i \neq j,$$

where

$$\rho(i, j) = \frac{\gamma(i, j)}{\sigma^2}$$

is the autocorrelation between X_i and X_j, and

$$\gamma(i, j) = E[(X_i - \mu)(X_j - \mu)]$$

is the autocovariance between X_i and X_j.

The questions one needs to answer are:

1. How realistic are these assumptions?

2. If one or more of these assumption does not hold, to what
   extent are (1.1), (1.2), and (1.3) wrong and how can they be
   corrected?

The first two assumptions depend on the marginal population
distribution F only. Here, our main concern is assumption 3. Unless specified otherwise, we therefore assume throughout the book
that the first two assumptions hold. The situation involving infinite
variance and/or mean is discussed briefly in Chapter 11.

Let us now consider assumption 3. In some cases this assumption is believed to be plausible a priori. In other cases, one tends to
believe that the dependence between the observations is so weak
that it is negligible for all practical purposes. In particular, in experimental situations one often hopes to force observations to be at
least approximately independent, by planning the experiment very
carefully. Unfortunately, there is ample practical evidence that this
wish does not always become a reality (see, e.g., Sections 1.4 and

1.5). A typical example is the series of standard weight measurements by the US National Bureau of Standards, which is discussed in Sections 1.4 and 7.3. This example illustrates that non-negligible persisting correlations may occur, in spite of all precautions. The reasons for such correlations are not always obvious. Some possible "physical" explanations are discussed in Section 1.3 (see also Section 1.5.5).

Let us now consider the question of what happens to (1.1) when the observations are correlated. For $\bar{X}_n$ to be meaningful, we assume that $E(X_t) = \mu$ is constant. The variance of $\bar{X} = n^{-1} \sum_{i=1}^{n} X_i$ is equal to

$$\text{var}(\bar{X}) = n^{-2} \sum_{i,j=1}^{n} \gamma(i,j) = n^{-2}\sigma^2 \sum_{i,j=1}^{n} \rho(i,j). \qquad (1.4)$$

If the correlations for $i \neq j$ add up to zero, then

$$\sum_{i,j=1}^{n} \gamma(i,j) = \sum_{i=1}^{n} \gamma(i,i) = \sum_{i=1}^{n} \sigma^2 = n\sigma^2.$$

This means that, if

$$\sum_{i \neq j}^{n} \rho(i,j) = 0, \qquad (1.5)$$

then (1.1) holds. In particular this is the case if $X_1, ..., X_n$ are mutually uncorrelated. Otherwise, if (1.5) does not hold, then the variance of $\bar{X}$ is equal to

$$\sigma^2[1 + \delta_n(\rho)]n^{-1} \qquad (1.6)$$

with the non-zero correction term

$$\delta_n(\rho) = n^{-1} \sum_{i \neq j} \rho(i,j). \qquad (1.7)$$

If the correlations $\rho(i,j)$ depend only on the lag $|i - j|$, then (1.7) can be simplified to

$$\delta_n(\rho) = 2 \sum_{k=1}^{n-1} (1 - \frac{k}{n})\rho(k). \qquad (1.8)$$

For simplicity we use the notation $\gamma(i - j) = \gamma(j - i) = \gamma(i,j)$ and $\rho(i - j) = \rho(j - i) = \rho(i,j)$. We will call a stochastic process with such correlations and a constant mean $\mu = E(X_i)$ stationary.

Table 1.1. *Correction factor $c(a)$ such that* $\text{var}(\bar{X}_n) = \sigma^2 c(a)n^{-1}$ *where* X_i *is an AR(1)-process* $X_i = aX_{i-1} + \epsilon_i$.

a	-0.8	-0.6	-0.4	-0.2	0	0.2	0.4	0.6	0.8
$c(a)$	0.11	0.25	0.43	0.67	1.00	1.50	2.33	4.00	9.00

   Consider, for example, an AR(1) process (first-order autoregressive process):

$$X_i = aX_{i-1} + \epsilon_i, \ a \in (-1, 1) \tag{1.9}$$

with independent increments ϵ_i that are normally distributed with zero mean and constant variance σ_ϵ^2. Then

$$\rho(i, j) = \rho(i - j) = a^{|i-j|}. \tag{1.10}$$

Combining (1.4), (1.8), and (1.10),

$$\text{var}(\bar{X}) = n^{-2} \sum_{i,j=1}^{n} \gamma(i, j) = n^{-2}\sigma^2 [\sum_{i=1}^{n} 1 + \sum_{i \neq j} a^{|i-j|}]$$

$$= \sigma^2 n^{-1}[1 + 2 \sum_{k=1}^{n-1} (1 - \frac{k}{n})a^k].$$

This can be written as

$$\sigma^2[1 + \delta_n(a)]n^{-1} = \sigma^2 c_n(a)n^{-1}, \tag{1.11}$$

with

$$\delta_n = \frac{2a}{1-a}[1 - n^{-1}\frac{1}{1-a} + n^{-1}\frac{a^n}{1-a}]. \tag{1.12}$$

Thus for large samples,

$$\text{var}(\bar{X}) \approx \sigma^2[1 + \delta(a)]n^{-1} = \sigma^2 c(a)n^{-1}, \tag{1.13}$$

with

$$\delta(a) = \frac{2a}{1-a}. \tag{1.14}$$

The right-hand side of (1.1) has to be multiplied by the constant $c(a) = 1 + \delta(a)$. How much the correction factor $c(a)$ differs from 1 depends on a. When a is close to 1, then $c(a)$ is large. In the limit we obtain $c(a) \to \infty$ for $a \to 1$. On the other hand, $c(a) \to 0$ for $a \to -1$. Table 1.1 displays $c(a)$ for different values of a.

   The numbers in Table 1.1 illustrate that, if the dependence is very strong ($|a|$ is close to 1), then $\sigma^2 n^{-1}$ is a very poor approximation of the variance of $\bar{X}_n$. In such a case it is important to have a good estimate of $c(a)$. Fortunately, a strong dependence between observations that are not far apart usually becomes quite obvious, by plotting the series against time. If a is very close to 1, then successive observations will clearly tend to assume similar values. Similarily, if a is close to -1, then successive observations will clearly tend to alternate their position with respect to an overall mean. The danger of not noticing that $|\delta(a)|$ is large is therefore not too extreme.

It is important to note that, apart from the constant $c(a)$, formula (1.1) does not need any modification. The variance of $\bar{X}$ is still proportional to n^{-1}. The same is true for any sequence of observations for which

$$\delta(\rho) = \lim_{n \to \infty} \delta_n(\rho) = \lim_{n \to \infty} n^{-1} \sum_{i \neq j} \rho(i,j) \qquad (1.15)$$

exists, is finite, and is larger than -1. Asymptotically we then have

$$\mathrm{var}(\bar{X}) \approx \sigma^2 [1 + \delta(\rho)] n^{-1} = \sigma^2 c(\rho) n^{-1}. \qquad (1.16)$$

Most time series models known in the literature exhibit this behavior. The best known are ARMA processes and Markov processes. For a thorough discussion of ARMA processes, their properties, and statistical applications, we refer the interested reader to standard time series books such as those by Box and Jenkins (1970), Priestley (1981), and Brockwell and Davis (1987).

Equation (1.16) is a generalization of (1.1) in that it allows the constant $c(\rho)$ to differ from 1. Is this generalization sufficient? In principle, the constant $c(\rho)$ can assume any positive value. One might therefore think that, for all practical purposes, (1.16) is flexible enough. It turns out however that for some data sets, the variance of $\bar{X}$ appears to differ from (1.1) not just by a constant factor but rather by the speed at which it converges to zero. Data examples that exhibit such behavior are discussed in Sections 1.4 and 1.5. Naturally, since one has only a finite number of observations, it is not possible to prove with absolute certainty that $\mathrm{var}(\bar{X})$ tends to zero slower than n^{-1}. For a given n there is always a constant c such that $\mathrm{var}(\bar{X}) = c \cdot n^{-1}$. For instance, we always can find an ARMA model with a suitable value of c. If, however, the actual data generating process is indeed such that the variance of $\bar{X}_n$ decays to zero slower than n^{-1}, then c is not constant but is increasing with n. Fitting the "best" ARMA model (best in some mathematically defined sense) will then lead to using ARMA-models with many parameters. With increasing sample size, the number of parameters will tend to infinity. For many reasons, using an excessive number of parameters is undesirable, especially because it increases the uncertainty of the statistical inference, and the parameters are difficult to interpret. Therefore, if there is an indication of a slower decay of the variance of $\bar{X}$, it is useful to model this decay explicitly. The simplest approach one can think of is a decay proportional to $n^{-\alpha}$ for some $\alpha \in (0, 1)$. In other words,

$$\text{var}(\bar{X}) \approx \sigma^2 c(\rho) n^{-\alpha} \tag{1.17}$$

for some constant $\alpha \in (0, 1)$ where $c(\rho)$ is defined by

$$c(\rho) = \lim_{n \to \infty} n^{\alpha-2} \sum_{i \neq j} \rho(i, j). \tag{1.18}$$

The connection between (1.17) and the correlation structure can be seen most easily by considering correlations that depend on the lag $|i-j|$ only. From (1.8) and (1.18) it follows that with increasing sample size n, the sum of all correlations with lags $-n + 1, -n + 2, ..., -1, 0, 1, 2, ..., n - 2, n - 1$ must be proportional to $n^{1-\alpha}$, i.e.,

$$\sum_{k=-(n-1)}^{n-1} \rho(k) \approx \text{constant} \cdot n^{1-\alpha}. \tag{1.19}$$

As α is less than 1, this implies that

$$\sum_{k=-\infty}^{\infty} \rho(k) = \infty. \tag{1.20}$$

Thus the correlations decay to zero so slowly that they are not summable. More specifically, (1.19) holds if

$$\rho(k) \approx c_\rho |k|^{-\alpha} \tag{1.21}$$

as $|k|$ tends to infinity and c_ρ is a finite positive constant. The intuitive interpretation of (1.21) is that the process has long memory. The dependence between events that are far apart diminishes very slowly with increasing distance. A stationary process with slowly decaying correlations (1.21) is therefore called a stationary process with long memory or long-range dependence or strong dependence (in contrast to processes with summable correlations which are also called processes with short memory or short-range correlations or weak dependence). We will see in Chapter 2 that equation (1.21) is essentially equivalent to the spectral density having a pole at zero. Recall that the spectral density f is defined by (see, e.g., Priestley

1981):

$$f(\lambda) = \frac{\sigma^2}{2\pi} \sum_{k=-\infty}^{\infty} \rho(k)e^{ik\lambda}. \tag{1.22}$$

Then (1.21) implies

$$f(\lambda) \approx c_f |k|^{\alpha-1}, \tag{1.23}$$

as $\lambda \to 0$ where c_f is a positive constant. Thus, for $\alpha < 1$, f tends to infinity at the origin. In contrast to long-range dependence, (1.15) is finite for processes with correlations that decay so quickly to zero that

$$\sum_{k=-\infty}^{\infty} \rho(k) = \text{constant} < \infty. \tag{1.24}$$

For example, for ARMA and Markov processes, the asymptotic decay of the correlations is exponential in the sense that there is an upper bound

$$|\rho(k)| \leq ba^k \tag{1.25}$$

where $0 < b < \infty$, $0 < a < 1$ are constants. Because the absolute value of a is less than 1, (1.24) holds. In the special case of an AR(1) process, this was previously illustrated explicitly.

It is important to note that equation (1.21) determines only the decay of the correlations. It does not say that there are some specific lags for which $\rho(k)$ is particularly large. This makes the detection of long-range dependence more difficult. We cannot just look at one or a few selected correlations to notice that (1.21) holds. We rather have to judge the way the correlations converge to zero with increasing lag. For statistical inference, it is the combined effect of all correlations that determines how accurate are the rules derived under independence. Therefore, in spite of being difficult to detect, long-range correlations are relevant for statistical inference, even if their individual values seem rather small. For instance, the effect of (1.21) on the variance of $\bar{X}$ can be extreme, even in the case of small individual correlations. What matters is that the sum of the correlations is large. In contrast to that, for short-range correlations with (1.24), the correction δ_n is typically made large by a few individually large correlations. For example, the sum of correlations of the form (1.10) can be only large when a is large. Because $\rho(k) = a^{|k|}$, this implies that the lag-1 correlation is large. We are therefore likely to detect the dependence by estimating $\rho(1)$ from the data. Strictly mathematically speaking, (1.25) also determines the asymptotic behavior of $\rho(k)$ only. However, it gives an upper bound that decays to zero very fast. If (1.25) does not hold for

Table 1.2. *Comparison of* $v_o = \sigma n^{-\frac{1}{2}}$ *with* $v_1 = \text{var}(\bar{X}_n)^{-\frac{1}{2}}$ *for* $\rho(k) =$ $a^{|k|}$ *and for* $\rho(k) = \gamma \cdot |k|^{-0.2}$ $(\gamma = 0.1, 0.5, 0.9)$. *Listed are the ratios* $q_n = v_1/v_o$. *Also given are the maximal correlations* $\rho_{max} = \max_k \rho(k)$.

	ρ_{max}	$n = 50$	$n = 100$	$n = 400$	$n = 1000$
$\rho(k) = a^{|k|}$					
$\quad a = 0.1$	0.1	1.108	1.107	1.106	1.106
$\quad a = 0.5$	0.5	1.755	1.744	1.735	1.733
$\quad a = 0.9$	0.9	4.752	4.561	4.410	4.380
$\rho(k) = 0.1 \cdot |k|^{-0.2}$	0.1	2.007	2.526	4.197	5.978
$\rho(k) = 0.5 \cdot |k|^{-0.2}$	0.5	4.018	5.283	9.169	13.218
$\rho(k) = 0.9 \cdot |k|^{-0.2}$	0.9	5.316	7.032	12.269	17.711

small lags, then this means that some (or even all) correlations for small lags are larger than this upper bound. This would make it even easier to detect the correlations, if b and a are large. If, on the other hand, a and b are small and all correlations are smaller or at most slightly larger than the upper bound, then $\delta_n(\rho)$ defined in (1.7) is small anyway. It is then practically not so important to detect the dependence. The effect of (1.21) and (1.25) on $\delta_n(\rho)$ as compared to the size of the correlations for small lags is illustrated in Table 1.2. For statistical inference, standard deviations are the relevant quantities. As examples we compare the standard deviation of $\bar{X}_n$ under independence with that for the AR(1) model (1.9) and a process with slowly decaying correlations $\rho(k) = \gamma \cdot |k|^{-0.2}$ $(\gamma = 0.1, 0.5, 0.9)$. A slow decay of the correlations at the approximate rate $k^{-0.2}$ is not uncommon in practice. A comparison with the maximal autocorrelation shows that, for the AR(1) process, the standard deviation of $\bar{X}_n$ differs substantially from $\sigma n^{-\frac{1}{2}}$, only if the lag-1 correlation is large. On the other hand, for the slowly decaying correlations, the effect on the standard deviation of the sample mean is large even if $\rho(k)$ never exceeds 0.1.

The effect of dependence on coverage probabilities is illustrated in Table 1.3. The factor by which (1.2) needs to be stretched is the same as the ratio q_n, namely

$$\sqrt{c_n(\rho)} = \sqrt{1 + \delta_n(\rho)}, \qquad (1.26)$$

where $\delta_n(\rho)$ is defined by (1.7). Asymptotically this is equal to the

Table 1.3. *Comparison of the coverage probability of* (1.2) *with (incorrect) nominal coverage probability 0.95. Observations are assumed to be normal with correlations as in Table 1.2. Also listed are the maximal correlations* $\rho_{max} = max_k \rho(k)$.

	ρ_{max}	$n = 50$	$n = 100$	$n = 400$	$n = 1000$
$\rho(k) = a^{|k|}$					
$\quad a = 0.1$	0.1	0.923	0.924	0.924	0.924
$\quad a = 0.5$	0.5	0.756	0.739	0.741	0.742
$\quad a = 0.9$	0.9	0.320	0.333	0.343	0.346
$\rho(k) = 0.1 \cdot |k|^{-0.2}$,	0.1	0.671	0.562	0.359	0.257
$\rho(k) = 0.5 \cdot |k|^{-0.2}$,	0.5	0.374	0.289	0.169	0.118
$\rho(k) = 0.9 \cdot |k|^{-0.2}$,	0.9	0.288	0.220	0.127	0.088

constant

$$\sqrt{c(\rho)} = \sqrt{1 + \delta(\rho)}, \qquad (1.27)$$

where $\delta(\rho)$ is as in (1.15), provided that this limit exists. If instead the correlations decay so slowly that for large sample sizes the variance of $\bar{X}$ is given by (1.17), then the situation is more dramatic. The confidence interval given by (1.2) is still incorrect by the factor $\sqrt{1 + \delta_n(\rho)}$. This time, however, $\delta_n(\rho)$ does not converge to a constant but diverges to plus infinity at the rate $n^{1-\alpha}$. Therefore, (1.23) has to be multiplied not just by a constant but by a constant times a positive power of n,

$$n^{\frac{1}{2}(1-\alpha)}\sqrt{c(\rho)}, \qquad (1.28)$$

where $c(\rho)$ is defined by (1.18). The results in Table 1.3 illustrate that the effect of slowly decaying correlations is dramatic, even for relatively small sample sizes. As the sample size increases, (1.2) and (1.3) become more and more unreliable.

An additional complication for correlated data is worth noting. Namely, s^2 is a biased estimator of σ^2. The bias depends directly on the correlation structure by

$$E(s^2) = \sigma^2[1 - (n-1)^{-1}\delta_n(\rho)], \qquad (1.29)$$

with $\delta_n(\rho)$ defined by (1.7). If the observations are uncorrelated, then $\delta_n(\rho) = 0$ and we obtain the well-known result that s^2 is unbiased. If the correlations are predominantly positive, then $\delta_n(\rho)$ is positive and the sample variance tends to underestimate σ^2.

Table 1.4. *Relative bias* $\Delta\sigma^2 = [E(s^2) - \sigma^2]/\sigma^2$ *of the sample variance*
s^2 *for the same models as in Table 1.2.*

	ρ_{max}	$n = 50$	$n = 100$	$n = 400$	$n = 1000$
$\rho(k) = a^{|k|}$					
$a = 0.1$	0.1	-0.0046	-0.0023	-0.0006	-0.0002
$a = 0.5$	0.5	-0.0424	-0.0206	-0.0050	-0.0020
$a = 0.9$	0.9	-0.4404	-0.2000	-0.0462	-0.0182
$\rho(k) = 0.1 \cdot |k|^{-0.2}$,	0.1	-0.0618	-0.0544	-0.0416	-0.0348
$\rho(k) = 0.5 \cdot |k|^{-0.2}$,	0.5	-0.3091	-0.2719	-0.2082	-0.1739
$\rho(k) = 0.9 \cdot |k|^{-0.2}$,	0.9	0.5564	-0.4893	-0.3748	-0.3130

Asymptotically the bias diappears. However, for small or moderately large sample sizes the negative bias makes the already existing underestimation of the uncertainty by (1.3) even greater. This
is illustrated in Table 1.4. If the correlations are predominantly
negative, then $\delta_n(\rho)$ is negative and the sample variance tends to
overestimate σ^2. Again, the bias makes our error in assessing uncertainty by (1.3) even greater, this time by overestimating the
uncertainty. It should also be noted that, if $\alpha < \frac{1}{2}$, then the variance of s^2 converges to zero slower than $1/n$ (See section 8.4). Thus,
s^2 is much less precise than in the case of short-memory processes.
Scale estimates that have the same rate of convergence as under
short-range dependence can be obtained, however, by maximum
likelihood estimation (see Chapters 5 and 6).

Finally, it should be noted that, we implicitly assumed the sample mean, appropriately normalized, to be asymptotically normally
distributed. In the presence of long-range correlations, this is not
necessarily true in general, even if all moments of X_i exist (see
Chapter 3). This complicates the issue of defining appropriate confidence intervals based on $\bar{X}_n$, though one might hope that in most
cases, asymptotic normality holds. Conditions under which this is
the case are discussed in chapter 3.

In this introduction we discussed only the very simple case where
$X_1, ..., X_n$ are identically distributed and we want to estimate
the expected value $\mu = E(X_i)$. We restricted attention to this
case, in order to illustrate the problem of slowly decaying correlations, without cluttering the exposition with cumbersome notation.
The same problem has to be addressed, however, for many other

statistical methods. For example, in a standard regression model $Y_i = \beta_o + \beta_1 x_{i,1} + \beta_2 x_{i,2} + ... + \beta_k x_{i,k} + \epsilon_i$ the errors ϵ_i are assumed to be independent. What happens when they are correlated and the asymptotic decay of the correlations follows (1.20)? Again, similar remarks as presented earlier apply though the situation is more complicated due to the presence of the explanatory variables $x_{i,j}$. This is discussed in Chapter 9.

1.2 Forecasting, an example

As we saw in the previous section, slowly decaying correlations make the estimation of constants, such as the mean, more difficult. The opposite is true for predictions of future observations. The more dependence there is between a future observation X_{n+k} and past values $X_n, X_{n-1}, X_{n-2}, ...$, the better X_{n+k} can be predicted, provided that the existing dependence structure is exploited appropriately. To obtain good forecasts, it is important to use an appropriate model. Consider for instance Figure 1.1.

It displays the yearly minimal water levels of the Nile River for the years 1007 to 1206, measured at the Roda Gauge near Cairo (Tousson, 1925, p. 366-385). This is a part of a longer series that is historically of particular interest. It is discussed in more detail in Section 1.4. In Figure 1.1, the series is split into two halves. Assuming that we observed the years 1007 to 1106, we predict the next 100 years. On one hand, we give optimal k-steps-ahead predictions based on an autoregressive process $X_t - \mu = \beta_1(X_{t-1} - \mu) + ... + \beta_p(X_{t-p} - \mu) + \epsilon_t$ with $\mu = E(X_t)$ and ϵ_t iid zero mean normal. The order p is obtained by Akaike's order selection criterion AIC (Akaike 1973a,b), using the *Splus* function *ar.burg*. The AIC is generally recommended when optimal prediction is the aim of the statistical analysis (see, e.g., Shibata 1980). Here, the chosen order turns out to be 2. On the other hand, the optimal predictions based on a so-called fractional ARIMA$(0, d, 0)$-process are given. The exact definition of this model is given in Chapter 2. The essential point is that it is a very simple process with slowly decaying correlations that are asymptotically proportional to k^{2d-1}. Its correlations are described by one parameter only, namely $d = H - 1/2$. Figure 1.1 illustrates the typical difference between classic Box-Jenkins forecasts and forecasts based on processes with long memory. The forecasts based on the short-memory process converge very fast, at an exponential rate, to the sample mean $\bar{X}_n$ of the past observations. Using the sample mean

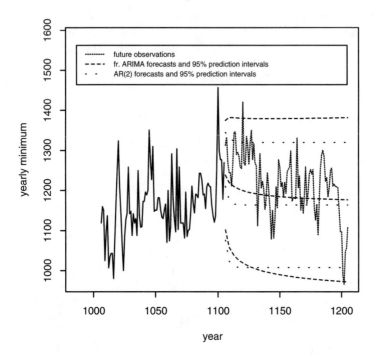

Figure 1.1. *Nile River minima: k-steps-ahead-forecasts based on the last 100 observations (k = 1, ..., 100)*

corresponds to total ignorance about the future observations (except for their unconditional expected value). On the other hand, the forecasts based on the process with slowly decaying correlations converge rather slowly to $\bar{X}_n$. This means that past observations influence the forecasts even far into the future. In the current example, this appears to be more appropriate. Also, the prediction intervals for the autoregressive process do not seem to have the desired coverage probability. Relatively many of the short-term forecasts are outside of the 95% prediction interval. This is also illustrated in Figures 1.2a and b, where we compare the absolute values of the observed k-steps-ahead prediction errors $\Delta_k(i) = |X_{1106+k} - \hat{X}(k, i)|$ $(i = 1, 2)$, where X_{1106+k} is the observation in the year $1106 + k$, $\hat{X}(k, 1)$ is its AR(2) forecast, and

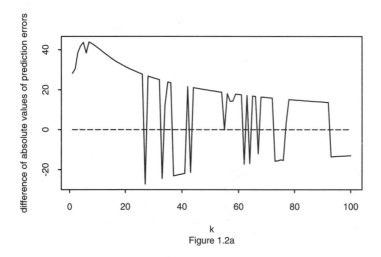

Figure 1.2a

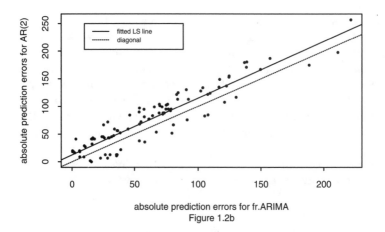

absolute prediction errors for fr.ARIMA
Figure 1.2b

Figure 1.2. *Nile River minima: plot of* $\Delta_k(1) - \Delta_k(2)$ *vs.* k *(Figure 1.2a) and* $\Delta_k(1)$ *vs.* $\Delta_k(2)$ *(Figure 1.2b), where* $\Delta_k(i)$ *is the absolute value of the observed k-steps-ahead prediction error for the AR(2) forecast ($i = 1$) and fr.ARIMA(0, d, 0) forecast ($i = 2$), respectively.*

$\hat{X}(k, 2)$ its fractional ARIMA$(0, d, 0)$ forecast. The plots of the difference $\Delta_k(1) - \Delta_k(2)$ (Figure 1.2a) and of $\Delta_k(1)$ versus $\Delta_k(2)$ (Figure 1.2b) show that the observed forecasting error is almost always larger for the short-memory model.

1.3 "Physical" models with long memory

1.3.1 General remarks

Stochastic processes may be used to model the behavior of an observed time series in a purely statistical way, without a direct physical interpretation of the parameters. For instance, we may use a parametric model, which is characterized by a parameter vector θ, to predict future observations or to obtain a confidence interval for the mean. For this purpose, it is not necessary to establish a direct link between the estimated parameter vector $\hat{\theta}$ and the actual data generating process. Nevertheless, a parsimonious model that can be explained by the "physical" mechanism that generates the data would be preferable for several reasons. It is justified by more than statistical analysis only and therefore more likely to be approximately correct. Estimated parameters have a direct physical interpretation. This makes it much easier to analyze observed data in a meaningful way and may give more useful insight to the applied scientist than would a purely descriptive model. Unfortunately, cases where "physical" models are known seem to be the exception. In the next four sections, we discuss four approaches that grew out of the attempt to understand the occurrence of long memory in certain data sets in economics, physics, textile engineering, and agronomy.

1.3.2 Aggregation of short-memory models

In some statistical applications, the observed time series is an aggregate of many individual time series. For instance, many economic data sets are generated in this way. Also the global temperature series, which is discussed in Section 1.4, is obtained by (spatial) aggregation.

The question arises as to whether aggregation has any relevant effect on the dependence structure. To be more specific, suppose that individual time series $X_t(j)$ $(j = 1, 2, 3...)$ are summarized in one single time series by

$$X_t = \sum_{j=1}^{\infty} X_t(j). \tag{1.30}$$

Assume further that each individual time series is stationary with short memory. Is it possible that the aggregated time series X_t exhibits long memory? There are at least two reasons why one is interested in this question:

1. If long memory can be induced artificially by averaging, then it might be better to build a model for the collection of individual time series $X_t(j)$ instead of analyzing the aggregated time series directly.

2. If long memory can be induced artificially by averaging, then this provides a possible explanation for long memory in an observed series. It might then be worth investigating whether one can find "physical" reasons for long memory other than aggregation.

This question was addressed by Granger (1980). It turns out that aggregation of short-memory processes can indeed lead to long memory. Granger considered the following aggregation model. Let $X_t(j)$, $j = 1, 2, \dots$ be independent AR(1) processes defined by

$$X_t(j) = \alpha_j X_{t-1}(j) + \epsilon_t(j), \tag{1.31}$$

where $-1 < \alpha_j < 1$, $\epsilon_1(j)$, $\epsilon_2(j), \dots$ are independent zero mean random variables with variance σ_j^2 and the processes $[\epsilon_t(1)]_{t\in N}$, $[\epsilon_t(2)]_{t\in N}, \dots$ are independent from each other. The spectral density of each individual series $X_t(j)$ is equal to

$$f(\lambda) = \frac{\sigma_j^2}{2\pi} \frac{1}{|1 - \alpha_j e^{-i\lambda}|^2}. \tag{1.32}$$

If we build the sum of N processes

$$X_t^{(N)} = X_t(1) + X_t(2) + \dots + X_t(N), \tag{1.33}$$

then $X_t^{(N)}$ has the spectral density:

$$f^{(N)}(\lambda) = \sum_{j=1}^{N} f_j(\lambda). \tag{1.34}$$

Suppose now that the parameters α_j and σ_j^2 are independent realizations of random variables U and V, drawn randomly from the population distributions F_α with mean α and F_σ with mean σ_ϵ^2, respectively. Also, assume that all $\alpha_1, \alpha_2, \ldots$ and $\sigma_1^2, \sigma_2^2, \ldots$ are independent. Then, for large N, the spectral density of $X_t^{(N)}$ can be approximated by the expected value

$$E[f_j(\lambda)] = \frac{N}{2\pi} E[V|1 - Ue^{-i\lambda}|^{-2}]. \qquad (1.35)$$

Because V and U are independent, we obtain

$$f^{(N)}(\lambda) \approx \frac{N}{2\pi}\sigma_\epsilon^2 E[|1 - Ue^{-i\lambda}|^{-2}] = \frac{N}{2\pi}\sigma_\epsilon^2 \int \frac{1}{|1 - ue^{-i\lambda}|^2} dF_\alpha(u). \qquad (1.36)$$

Granger showed that if F_α is a beta distribution with suitable pararameters, then the right-hand side of (1.36) has an integrable pole at zero. Thus, the aggregated series $X_t^{(N)}$ approaches a limiting time series X_t [equation (1.30)] that has long memory, although each individual time series from which it is built is a simple AR(1) process. More complicated models can be considered if necessary (see, e.g., Granger's remarks on dependent series, micro-models, feedback, etc.). The main message is that from observing long-range dependence in an aggregated series (*macro-level*), one cannot necessarily conclude that this long memory is due to the genuine occurrence of long memory in the individual series (*micro-level*). It is possible that instead it is induced artificially by aggregation. To find the source of long memory, the behavior of the individual series needs to be examined. This might be possible in some cases, whereas in other cases individual series might not be available.

1.3.3 Critical phenomena

An introduction to the role of long memory in the context of critical phenomena in physics is given in Cassandro and Jona-Lasinio (1978). Thermodynamic systems can be studied by considering random fields on a finite but large d-dimensional lattice $\Lambda \subset Z^d$. For instance, in the so-called Ising model, each point $i \in \Lambda$ is associated with a sign,

$$S_i \in \{+1, -1\}, \qquad (1.37)$$

representing the spin at this position in a ferromagnetic substance. In another interpretation, S_i could represent the presence of a gas molecule at location i of a "lattice gas"; $S_i = 1$ stands for "molecule present" and $S_i = -1$ stands for "molecule absent". The energy of a configuration,

$$S(\Lambda) = \{S_i : i \in \Lambda\}, \qquad (1.38)$$

is given by the (finite volume) Hamiltonian

$$H_\Lambda(S) = - \sum_{i,j \in \Lambda} J_{ij} S_i S_j - h \sum_{i \in \Lambda} S_i. \qquad (1.39)$$

Here J_{ij} are parameters determining the type of interaction between neighboring spins. For instance, if $J_{ij} = J_{|i-j|}$ and $J_1 > 0$, then neighboring spins tend to be aligned in the same direction. Alignment in the same direction is typical for a ferromagnetic substance. $M_\Lambda(S) = \sum_{i \in \Lambda} S_i$ is called total magnetization; the product $h M_\Lambda(S)$ characterizes the interaction with an external magnetic field. Physical properties of the system are studied by considering the thermodynamic limit $\Lambda \to Z^d$. In particular, one needs to find a suitable normalization $\sigma(\Lambda)$ such that

$$M_\Lambda^*(S) = \frac{M_\Lambda(S) - E[M_\Lambda(S)]}{\sigma(\Lambda)} \qquad (1.40)$$

converges to a nondegenerate random variable. Under "normal conditions", the spins S_i are at most weakly dependent variables, so that

$$\sigma(\Lambda) \approx c \cdot |\Lambda|^{\frac{1}{2}}, \qquad (1.41)$$

where $|\Lambda|$ is equal to the number of elements in Λ and $c > 0$. However, there exists a critical temperature such that one observes correlations $\rho(i - j) = \mathrm{corr}(S_i, S_j)$ which decay slowly to zero as the Euclidian distance $||i - j||$ tends to infinity (see Cassandro and Jona-Lasinio 1978). This implies that

$$\sigma(\Lambda) \approx c \cdot |\Lambda|^\alpha \qquad (1.42)$$

for some $\frac{1}{2} < \alpha < 1$, $c > 0$. This situation is also called phase transition. For ferromagnetic substances it corresponds to spontaneous magnetization. For a gas, it is a model for the transition from a liquid phase to a gaseous phase.

Other areas in physics where long-range dependence occurs include for example, turbulence theory, quantum field theory, and $1/f$ noises (see, e.g., Cassandra and Jona-Lasinio 1978, Eberhard and Horn 1978, Marinari et al. 1983), in particular, the investigation of turbulences motivated Kolmogorov (1940) to define fractional Brownian motion and some related processes.

1.3.4 Hierarchical variation

In an interlaboratory standardization experiment, Cox and Townsend (1947) considered the coefficient of variation of mass per unit length of worsted yarn. Two hundred randomly chosen sections of lengths ranging from 1 to 10^4 cm were weighed. For lengths longer than the fiber length, the autocorrelations were estimated as a function of the distance k along the yarn. They were found to be proportional to $k^{-\alpha}$ for some $0 < \alpha < 1$.

Cox (1984) proposed a physical explanation for these correlations, by constructing a model with hierarchical variation. The textile yarns are produced by a repeated process of thinning and pulling out. Each time this process is applied, the current one-dimensional spatial scale is stretched due to the "pulling". In addition, new short-term variation ("noise") is introduced. Let $X_t^{(N)}$ denote the weight per unit length of the yarn at location t, after N stages of processing. Also, denote by $U_t^{(N)}$ the variation introduced at the N-th stage and let a be a constant attenuation factor. The processes $U^{(o)}(t), U^{(1)}(t), ...$ are assumed to be independent. A plausible model for $X_t^{(N)}$ is

$$X_t^{(N)} = \sum_{s=0}^{N} a^{-(N-s)-sb} U^{(s)}(ta^{-(N-s)}), \qquad (1.43)$$

where $0 < b < 1$ is some constant. Under mild regularity conditions on the autocorrelations of $U^{(N)}(t)$, it can be shown that for large N the covariances of $X_t^{(N)}$ at large lags k are proportional to $k^{-2(1-b)}$.

Although this is an example in a very specific context, it is a special case of hierarchical models that are also useful in physics (see, e.g., Cassandro and Jona-Lasinio 1978).

1.3.5 Partial differential equations

Using stochastic partial differential equations, Whittle (1956, 1962) gave a partial "physical" explanation for empirical findings by Smith (1938; see Section 1.5), which suggest a slow hyperbolic decay of spatial correlations in uniformity trials. Whittle considered a stochastic process Y_ν with $\nu = (x, t) = (x_1, ..., x_m, t)$ $m \geq 1$. The index t denotes time. More specifically, in the context of agricultural uniformity trials, Y_ν may represent "soil fertility". The soil is three dimensional, thus $m = 3$. There are two opposite effects on fertility. On one hand, there is a uniformizing effect due to diffusion of nutrient salts. On the other hand, there is a "disturbing" effect ϵ_ν due to weather, cultivation, artificial fertilization, etc. Ultimately, a stationary equilibrium may be reached between these two effects. These considerations led Whittle to the stochastic partial differential equation:

$$\frac{\partial Y_\nu}{\partial t} + \alpha Y_\nu = \frac{1}{2} \nabla^2 Y_\nu + \epsilon_\nu. \tag{1.44}$$

Here

$$\nabla^2 = \frac{\partial^2}{\partial x_1^2} + ... + \frac{\partial^2}{\partial x_m^2}$$

and ϵ_ν is a suitably chosen random field. The term $\nabla^2 Y_\nu$ represents spreading of fertility through the soil and αY_ν stands for a "direct loss" of fertility. Under suitable conditions on the "disturbing" process, Whittle showed that for $m > 2$, the solution of (1.44) is a stationary random field with spatial covariance function

$$\text{cov}(Y_{x,t}, Y_{x',t}) = \gamma(||x - x'||) \sim c \cdot ||x - x'||^{2-m}, \tag{1.45}$$

as $||x - x'|| \to \infty$, where $c > 0$ and $||.||$ denotes the Euclidian distance in R^m. Whittle also obtained several other power laws for the spatial covariance, by imposing various correlation structures on ϵ_ν and boundary conditions for Y_ν.

Some remarks on how to model spatial long-memory models via partial differential equations are given in Gay and Heyde (1990). Renshaw (1994) obtained general power laws for a one-dimensional "space" time model Y_ν, ($\nu = (i,t), i \in Z, t \geq 0$), space being one-dimensional. The model is defined in an intuitive way by

$$Y_{i,t+dt} - Y_{i,t} = \lambda Y_{i,t} dt + \sum_{j=-\infty}^{\infty} a_j (Y_{i,t} - Y_{i+j,t}) dt + dZ_{i,t} + o(dt),$$

$$\tag{1.46}$$

with $\lambda \geq 0$, $\sum_{j=-\infty}^{\infty} a_j < \infty$, and $dZ_{i,t}$ a white noise sequence of independent identically distributed random variables with mean 0 and variance $\sigma^2 dt$. The parameter λ represents the average constant growth rate of an individual at any given location i. The growth at location i is influenced by linear interactions of $X_{i,t}$ with individuals at other locations. Renshaw showed that, for certain choices of the interaction parameters a_j, the process reaches a stationary equilibrium with a spatial spectral density that has an integrable pole at the origin. Renshaw also obtained some spectra with a nonintegrable pole at zero. Such spectra seem to occur frequently in engineering, geophysics, and physics (see, e.g., Kolmogorov 1940: turbulences in a fluid; Akaike 1960: measurements of the roughness of roads and runways; Sayles and Thomas 1978: surface topographies; Sarpkaya and Isaacson 1981: sea waves on offshore structures; Bretschneider 1959: wind-generated waves). For

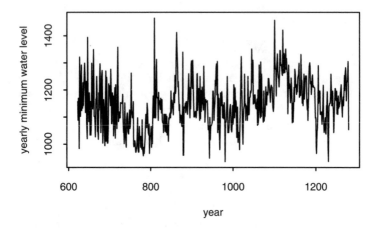

Figure 1.3. *Yearly minimum water levels of the Nile River at the Roda Gauge (622-1281 A.D.).*

a theoretical discussion of nonintegrable spectral densities see also Solo (1989a) and Matheron (1973).

1.4 Some data examples

We introduce five different data sets that will be used later to illustrate the theoretical results. Here, we briefly examine the dependence structure of these time series by considering plots (in log-log coordinates) of the sample autocorrelations against the lag, the variance of $\bar{X}_n$ against n, and the periodogram

$$I(\lambda_j) = \frac{1}{2\pi n} | \sum_{t=1}^{n} (X_t - \bar{X}_n) e^{it\lambda_j} |^2 = \frac{1}{2\pi} \sum_{k=-(n-1)}^{n-1} \hat{\gamma}(k) e^{ik\lambda_j} \quad (1.47)$$

at Fourier frequencies $\lambda_j = 2\pi j/n$ ($j = 1, ..., n^*; n^* =$integer part of $n/2 - 1/2$) against the frequency λ_j. Here $\hat{\gamma}(k)$ are the sample covariances

$$\hat{\gamma}(k) = \frac{1}{n} \sum_{t=1}^{n-|k|} (X_t - \bar{X}_n)(X_{t+|k|} - \bar{X}_n) \quad (1.48)$$

Series : x

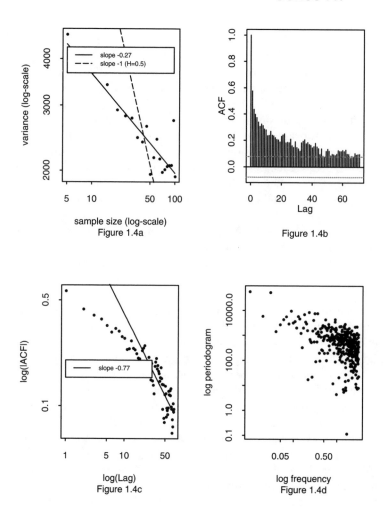

Figure 1.4. *Nile River minima: variance of sample mean vs. sample size on log-log-scale (figure 1.4a); sample autocorrelations (Figure 1.4b; on log-log scale in Figure 1.4c); periodogram on log-log scale (Figure 1.4d).*

for $|k| \leq n-1$ and by 0 for $|k| \geq n$. The periodogram is the sample analogon of the spectral density (1.22) in that the covariances $\gamma(k)$ are replaced $\hat{\gamma}(k)$.

Data set 1: Figure 1.3 displays the yearly minimal water levels of the Nile River for the years 622-1281, measured at the Roda Gauge near Cairo (Tousson, 1925, pp. 366-385). Historically, this data set is of particular interest. The analysis of this and several similar time series led to the discovery of the so-called Hurst effect (Hurst 1951). Motivated by Hurst's empirical findings, Mandelbrot and co-workers later introduced fractional Gaussian noise as a statistical model with long memory (see, e.g., Mandelbrot and Wallis 1968a,b 1969a,b,c, Mandelbrot and van Ness 1968). The presence of long memory in the Nile River minima can be seen in Figure 1.4a. It indicates that the variance of $\bar{X}_n$ converges to zero at a slower rate than n^{-1}. The fitted least squares line is $\log[\text{var}(\bar{X}_n)] = 8.82 - 0.27 \log(n)$. The slope -0.27 is far from -1 which is the theoretical value for summable correlations. Similarly, the plot of the autocorrelations in log-log coordinates suggests a slow decay of the correlations (Figure 1.4c). Further evidence for this type of long-range dependence is given by the log-log plot of the periodogram in Figure 1.4d. We will see later that, if the correlations were summable, then near the origin the periodogram should be scattered randomly around a constant. Instead, the points in Figure 1.4d are scattered around a negative slope. This is typically the case for stationary processes with nonsummable correlations.

Data set 2: Figure 1.5 displays the logarithm of the amount of coded information per frame for a certain video scene. This data set was provided by H. Heeke and E. Hundt (Siemens, Munich) through W. Willinger and A. Tabatabai (Bellcore, Morristown). The scene consists of a conversation among 3 people sitting at a table. There is no change in the background and no camera movement. The data set in Figure 1.6a (1000 frames) is part of a longer series based on about 30 minutes of video film. About 25 frames per second are processed. Fast transmission of digitally coded video pictures is an important issue for telecommunications companies. The data set here was generated by engineers at Siemens, Munich (see Heeke 1991, Heyman et al. 1991). A so-called variable-bit-rate (VBR) codec was used. This codec is especially designed for high-speed networks. The amount of information transmitted per time unit is adjusted to the momentary complexity of the picture. A characterization of the dependence structure of the information per frame can be useful for the assessment of the capacity

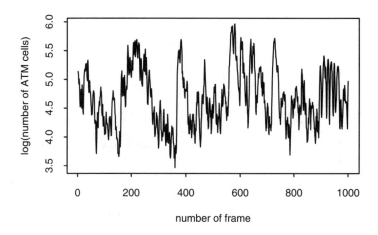

Figure 1.5. *VBR data: logarithm of the number of ATM cells per frame.*

a communication network has to have in order to guarantee reliable transmission. Figures 1.6a to 1.6c indicate that there are strong and long-lasting correlations. The estimated slope in Figure 1.6a is -0.53. The dependence structure, however, seems to be more complicated than for the Nile River data. Fitting a straight line in Figure 1.6c does not seem meaningful. The plot of the periodogram confirms the more complex nature of the dependence (Figure 1.6d). The slope around which the points in Figure 1.6d are scattered near the origin appears to be flatter than that for points farther away from the origin.

Data set 3: This data set consists of the number of bytes per millisecond, transformed by $\log(x + 1)$, which were sent through an Ethernet network (Figure 1.7). This is part of a large number of high-resolution Ethernet measurements for a local area network (LAN) at Bellcore, Morristown. The data were provided by W. Willinger, W. Leland, and D. Wilson (Bellcore, Morristown). Technical details are given in Leland and Wilson (1991); see also Fowler and Leland (1991) and Leland et al. (1993). The practical motivation for considering this type of data is essentially the same as for the VBR data above. Figure 1.7 is somewhat difficult to interpret, because of the partially discrete nature of the process. Figures 1.8b

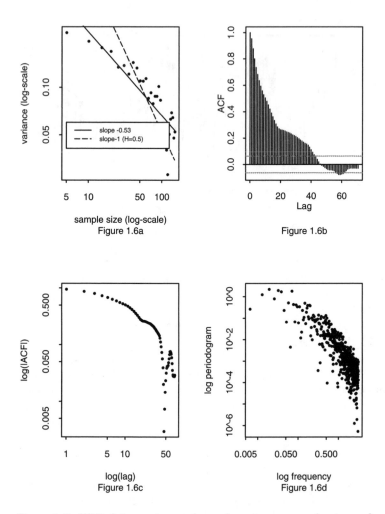

Figure 1.6. *VBR data: variance of sample mean vs. sample size on log-log scale (Figure 1.6a); sample autocorrelations (Figure 1.6b; on log-log scale in figure 1.6c); periodogram on log-log-scale (Figure 1.6d).*

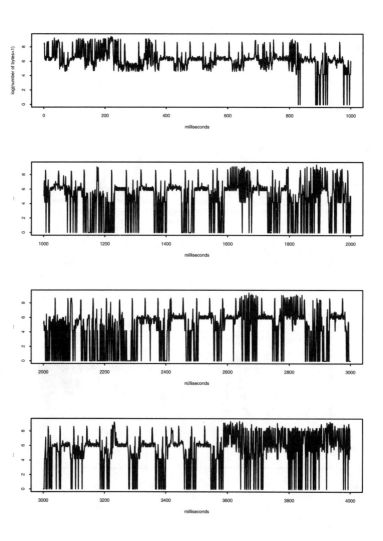

Figure 1.7. *Ethernet data:* $\log(x + 1)$ *where* x = *number of bytes.*

Series : x

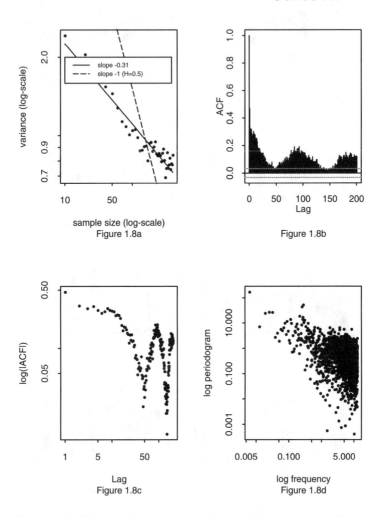

Figure 1.8. *Ethernet data: variance of sample mean vs. sample size on log-log-scale (Figure 1.8a); sample autocorrelations (Figure 1.8b; on log-log scale in Figure 1.8c); periodogram on log-log scale (Figure 1.8d)*

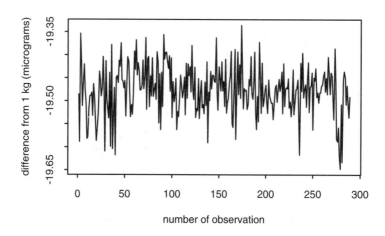

Figure 1.9. *NBS precision measurements on a 1 kg check standard weight (difference from 1 kg).*

and 1.8c, however, indicate clearly that the correlations are strong even for large lags. The fitted least squares slope in Figure 1.8a is -0.31. This impression is confirmed by the log-log plot of the periodogram (Figure 1.8d).

Data set 4: Figure 1.9 displays a series of high precision weight measurements of a 1kg check standard weight performed at the National Institute of Standards and Technology (NIST; formerly NBS), Gaithersburg, USA (see Pollak, Croarkin, and Hagwood 1993, Graf, Hampel, and Tacier 1984, Graf 1983). The data were provided by R.N. Varner (NIST). Additional information on the generation of the data was provided to me by C. Croarkin (NIST) and H.P. Graf (CIBA Geigy, Basel). The measurements were taken between June 24, 1963 and October 17, 1975, using the same weighing machine. The differences (in micrograms) from 1 kg were recorded. One difficulty with this data set is that the dates at which measurements were taken are not equidistant. This makes the analysis of short-term correlations problematic. However, the overall long-term dependence structure is likely to be almost unaffected. The estimated slope in the plot of the variance of $\bar{X}_n$ is -0.80 (Figure 1.10a). Based on this and Figures 1.10b and 1.10c,

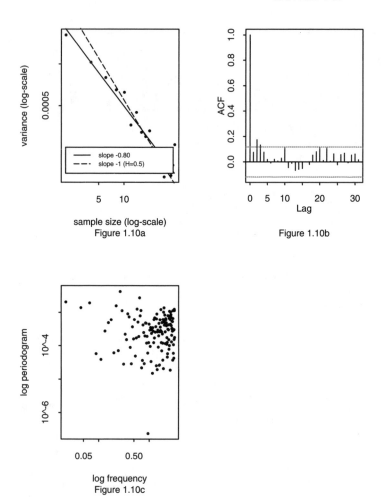

Figure 1.10. *NBS data: variance of sample mean vs. sample size on log-log scale (Figure 1.10a); sample autocorrelations (Figure 1.10b); periodogram on log-log scale (Figure 1.10c)*

one may or may not suspect the presence of slowly decaying correlations. The decision is much less clear-cut than in the examples
above. Nevertheless, we will see later (Section 7.3) that there is
some evidence for a significant departure from the hypothesis of
summable correlations.

Data set 5: Figure 1.11a displays the monthly temperature for
the northern hemisphere for the years 1854-1989, from the data
base held at the Climate Research Unit of the University of East
Anglia, Norwich, England (Jones and Briffa 1992). This data
set was provided by P.D. Jones (University of East Anglia) and
Richard Smith (University of North Carolina at Chapel Hill). The
numbers consist of the temperature (degrees C) difference from
the monthly average over the period 1950-1979. The plot of the
time series suggests an increasing trend that may reflect a "global
warming" during about the last 100 years. Figure 1.11b displays
the data after subtraction of an estimated linear trend. Apart from
a slightly larger variance at the beginnig, the residuals look more or
less stationary. Figures 1.12a to 1.12c suggest that the correlations
decay to zero rather slowly. The estimated slope in Figure 1.12a
is -0.45 which is rather far from -1. Moreover, although the data
were reportedly adjusted for seasonal effects, some seasonality still
seems to be present.

1.5 Other data examples, historic overview, discussion

1.5.1 Two types of situations

The examples discussed in the previous section can be divided into
two categories. For the precision measurements of the check standard weight, the occurrence of dependence was rather unsuspected.
For the other examples, dependence between the observations had
to be expected. In general, we can distinguish between two situations:

Situation 1: Due to the nature of the observed phenomenon
and/or the way observations are taken, dependence beween the
observations is expected a priori.

Situation 2: The observations are expected to be (more or less)
independent.

Naturally, sometimes one is somewhere in between these two extremes, since one might simply not be sure whether dependence
has to be expected or not. Typical examples for situation 1 are
cases where one cannot plan an experiment but rather makes ob-

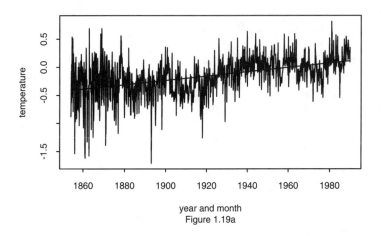

year and month
Figure 1.19a

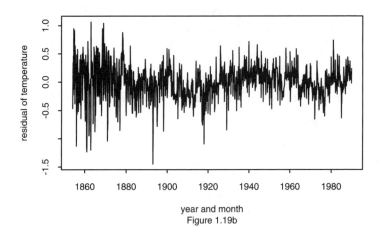

year and month
Figure 1.19b

Figure 1.11. *Global temperature for the northern hemisphere (1854-1989) (Figure 1.11a) and the residuals after subtraction of a linear trend (Figure 1.11b).*

Series : residual

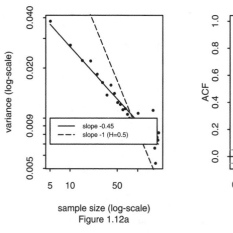

Figure 1.12a

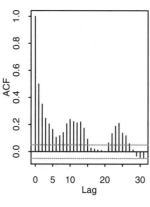

Figure 1.12b

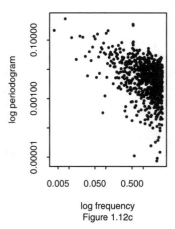

Figure 1.12c

Figure 1.12. *Detrended global temperature: variance of sample mean vs. sample size on log-log scale (Figure 1.12a); sample autocorrelations (Figure 1.12b); periodogram on log-log scale (Figure 1.12c)*

servations as they "come in". This is, for instance, often the case in areas like hydrology, geophysics, climatology, economics and agronomy. For example, the Nile River minima discussed in the previous section are not obtained from a controlled experiment. Instead, one observes the natural development of the river. A hydrologist would hardly expect successive observations to be independent. A complicated physical mechanism creates dependence between the data. Many known and unknown factors affect the water level. In contrast to an experimental situation, all or most of these influences cannot be controlled. The same is true for the ethernet data and the global temperature data, whereas the situation is somewhat different for the video data. There, the setup was experimental in the sense that the type of recorded scene was specified a priori. However, apart from that, no attempt was made to obtain independent observations. On the contrary, studying the dependence structure of the emerging code is one of the main aims. Typical examples for Situation 2 are experimental situations where one tries to eliminate any factor that could cause the observations to be dependent. This is, for instance, the case for the NBS check standard weights, where, it was intended to create an ideal experimental situation where measurement errors are independent.

The data sets in the previous section illustrated that correlations (expected or unexpected) not only occur, but they also may persist for a long time. This was recognized by many prominent applied statisticians and scientists many decades ago. In the following sections, we give a short overview on some of the important early references. This will also give rise to some principal considerations on the topic of long-range dependence, before going into the details of mathematical modelling.

1.5.2 The Joseph effect and the Hurst effect

Since ancient times, the Nile River has been known for its characteristic long-term behavior. Long periods of dryness were followed by long periods of yearly returning floods. Floods had the effect of fertilizing the soil so that in flood years the yield of crop was particularly abundant. On a speculative basis, one may find an early qualitative account of this in the Bible (Genesis 41, 29-30): "Seven years of great abundance are coming throughout the land of Egypt, but seven years of famine will follow them." We do not have any records of the water level of the Nile from those times. However, there are reasonably reliable historical records going as far back as

622 A.D. A data set for the years 622-1281 was discussed in the previous section. It exhibits a long-term behavior that might give an "explanation" of the seven "good" years and seven "bad" years described in Genesis. There were long periods where the maximal level tended to stay high. On the other hand, there were long periods with low levels. Overall, the series looks stationary. When one only looks at short time periods, then there seem to be cycles or local trends. However, looking at the whole series, there is no apparent persisting cycle. It rather seems that cycles of (almost) all frequencies occur, superimposed and in random sequence. Also, there is no global trend. In reference to the biblical "seven years of great abundance" and "seven years of famine," Mandelbrot called this behavior the *Joseph effect* (Mandelbrot 1977, 1983a, Mandelbrot and Wallis 1968a,b, Mandelbrot and van Ness, 1968).

The famous hydrologist Hurst (1951) noticed these characteristics when he was investigating the question of how to regularize the flow of the Nile River. More specifically, his discovery can be described as follows: Suppose we want to calculate the capacity of a reservoir such that it is ideal for the time span between t and $t + k$. To simplify matters, assume that time is discrete and that there are no storage losses (caused by evaporation, leakage, etc.). By ideal capacity we mean that we want to achieve the following: that the outflow is uniform, that at time $t+k$ the reservoir is as full as at time t, and that the reservoir never overflows. Let X_i denote the inflow at time i and $Y_j = \sum_{i=1}^{j} X_i$ the cumulative inflow up to time j. Then the ideal capacity can be shown to be equal to

$$R(t, k) = \max_{0 \le i \le k} [Y_{t+i} - Y_t - \frac{i}{k}(Y_{t+k} - Y_t)]$$

$$- \min_{0 \le i \le k} [Y_{t+i} - Y_t - \frac{i}{k}(Y_{t+k} - Y_t)] \qquad (1.49)$$

$R(t, k)$ is called the *adjusted range*. In order to study the properties that are independent of the scale, $R(t, k)$ is standardized by

$$S(t, k) = \sqrt{k^{-1} \sum_{i=t+1}^{t+k} (X_i - \bar{X}_{t,k})^2}, \qquad (1.50)$$

where $\bar{X}_{t,k} = k^{-1} \sum_{i=t+1}^{t+k} X_i$. Note that $S^2(t, k)$ is equal to $(k - 1)/k$ times the usual sample variance of $X_{t+1}, ..., X_{t+k}$. The ratio

$$R/S = \frac{R(t, k)}{S(t, k)} \qquad (1.51)$$

is called the *rescaled adjusted range* or R/S-statistic. Hurst plotted
the logarithm of R/S against several values of k. He observed that,
for large values of k, $\log R/S$ was scattered around a straight line
with a slope that exceeded $\frac{1}{2}$. In probabilistic terminology this
means that for large k,

$$\log E[R/S] \approx a + H \log k, \text{ with } H > \frac{1}{2}. \qquad (1.52)$$

This empirical finding was in contradiction to results for Markov
processes, mixing processes, and other stochastic processes that
where usually considered at that time. For any stationary process
with short-range dependence, R/S should behave asymptotically
like a constant times $k^{\frac{1}{2}}$. Therefore, for large values of k, $\log R/S$
should be randomly scattered around a straight line with slope
$1/2$. Hurst's finding that for the Nile River data, and for many
other hydrological, geophysical, and climatological records, R/S
behaves like a constant times k^H for some $H > \frac{1}{2}$, is known under
the name *Hurst effect*. Mandelbrot and co-workers showed that the
Hurst effect can be modelled by so-called fractional Gaussian noise
with self-similarity parameter $H \in (\frac{1}{2}, 1)$ (H for "Hurst"). We will
see in Chapter 2 how this process is defined exactly, and why H is
called a self-similarity parameter. For now, it is sufficient to state
that this model is a stationary process for which (1.21) holds.

1.5.3 Uniformity trials

Although we will focus on time series, it is worth noting that there
are interesting examples of long memory in spatial data. The results on uniformity trials obtained by Smith (1938) are famous.
Uniformity trials are a common method in agronomy to determine
the "best" size of a plot. By best we mean a size such that the
efficiency of a field experiment is optimized. Smith's definition of
efficiency incorporates cost and the average variability per unit
area. Basically, the aim is to obtain maximal information for the
lowest cost. In order to be able to assess the average variability,
Smith investigated the behavior of the sample mean as a function
of plot size. The following experiment was carried out: An experimental area was planted uniformly with wheat, in 38 rows, 6 inches
apart. Disregarding 4 rows at each side and 1 ft at each end of a
row (to avoid border effects), the remaining area of 15 ft by 36
ft (30 rows) was divided into 1080 elementary plots of $\frac{1}{2}$ square
ft. For each elementary plot the yield was recorded. For k varying

between 1 and 120, adjacent elementary plots were combined into nonoverlapping rectangular plots. For each of the resulting plots, the average yield was calculated. Finally, the variance $V(k)$ of the average yield for a plot of size k was estimated by the sample variance of the averages for plots of size k. Plotting $\log V(k)$ versus $\log k$, Smith found the relationship

$$\log V(k) = a + b \log k, \tag{1.53}$$

with b around -0.749. This implies that $V(k)$ converges to zero at a slower rate of convergence than if the observations were independent or weakly dependent. Smith reported the same results for a large number of other uniformity trials. In fact, he found the same behavior for all data he could find at that time.

Which dependence structure can explain (1.30)? Here we are in a slightly more difficult situation than in 1.3.2. The observations X_t are not ordered in time. The index $t \in N \times N$ gives the position in the plane. In principle, for each pair of positions t, s there is a separate correlation $\rho(t, s)$. In order to make the problem tractable, Whittle (1956) suggested assuming that $\rho(t, s)$ depends on the Euclidian distance between t and s only. He showed that if $\rho(t, s) = \rho(|t - s|)$ decays asymptotically like $|t - s|^{4H-4}$ for some $H \in (1/2, 1)$, then $V(k)$ converges to zero like a constant times k^{2H-2}. On the other hand, if the observations for different plots are independent of each other or if the correlations decay fast, then $V(k)$ asymptotically decays like a constant times k^{-1}. In a uniformity trial we would then expect to observe a value of b in the neighborhood of -1. Instead, in all uniformity trials considered by Smith, b was larger that -1. Clearly, dependence between neighboring plots has to be expected. It is the persistence of the dependence that is remarkable. Long-range dependence seems to be the rule rather than the exception in uniformity trials, and thus in most agricultural experiments. Fortunately, we will see later that, up to a certain degree, the undesired effects of such dependence can be eliminated by appropriate randomization.

1.5.4 Economic time series

Apart from Mandelbrot's pioneering work on so-called self-similar processes and their diverse applications, the importance of long-range dependence in ecomomic data was recognized by Granger (1966) in his article on *"The typical spectral shape of an economic variable."* Using different kinds of estimates of the spectral density, Granger observed that for economic time series, the typical shape of the spectral density is (at least in good approximation) a function with a pole at the origin. This is the case even after "known" business cycles and trends are removed. He formulated a qualitative "law": *"The long-term fluctuations in economic variables, if decomposed into frequency components, are such that the amplitudes of the components decrease smoothly with decreasing period."* Also, *"the same basic shape is found regardless of the length of the data available."* In a later article Granger (1980) suggested a "physical" explanation of how long-range dependence modelled by (1.21) may be caused by aggregation of many dynamic micro-relationships (see Section 1.3.2).

1.5.5 Semisystematic errors, unsuspected slowly decaying correlations, the personal equation

In the cases considered in Sections 1.5.2, 1.5.3 and 1.5.4, dependence (though not necessarily long-range dependence) between the observations was expected a priori. In other situations, everything seems well "under control" so that one does not expect any (or almost any) correlations. An interesting reference is Student (1927). Student discussed the source and nature of errors of routine analysis, with particular emphasis on chemical measurements. One of the main issues is a *"phenomenon which will be familiar to those who have had astronomical experience, namely that analyses made alongside one another tend to have similar errors; not only so but such errors, which I may call semi-constant, tend to persist throughout the day and some of them throughout the week or the month."* By *"semi-constant error"* he does not mean any systematic error or trend that could be explained or corrected easily: *"Why this is so is often quite obscure, though a statistical examination may enable the head of the laboratory to clear up large sources of error of this kind: it is not likely that he will eliminate all such errors."* For the chemist he therefore has interesting practical advise: *"The chemist who wishes to impress his clients will therefore arrange to do repetition analyses as nearly as possible at the same time, but if he wishes to diminish his real error he will separate them by as wide an interval of time as possible."* According to Student, the occurrence of "semi-constant errors" is the rule rather than the exception. Often no apparent reason can be found even after detailed investigation. These statements are illustrated by several data examples from chemical practice. One is the measurement of nitrogen in pure crystalline aspartic acid. Student described the experimental situation to be as much under control as one can possibly expect. The phenomenon of what Student calls "semi-constant errors" that *"tend to persist throughout the day and some of them throughout the week or the month"* is characteristic for stationary processes with slowly decaying correlations (1.21).

"Semi-constant errors" were known in astronomy before the publication of Student's article. The astronomer Newcombe (1895) observed that in astronomy, *"semi-systematic errors"* affect whole groups of observations, or in other words, successive observations tend to be on the same level for a long time. He concluded that this drastically increases the "probable error," so that the traditional standard error $\sigma/\sqrt{n}$ is clearly too small.

K. Pearson carried out experiments to simulate astronomical observations. The results are described in detail in his remarkable article "On the Mathematical Theory of Errors of Judgement, with Special Reference to the Personal Equation" in the Proceedings of the Royal Philosophical Society (1902). It is interesting to look at his results in more detail.

Two experiments were carried out:

Experiment 1: The same set of 500 lines with varying lengths was given to three independent observers (including Karl Pearson himself). Each observer bisected each of the 500 lines with a pencil stroke at sight. The judgments were made in the same room under the same circumstances.

Experiment 2: A second series of experiments was designed to *"test simulataneously the eye, the ear and the hand, and thus give every opportunity for a variety of small causes to influence the errors of judgment."* A machine was constructed such that the following experiment could be carried out: A very narrow beam of light traverses a white strip. At some instant a bell rings and the observer has to observe the position of the beam on the strip. Afterwards, he has to divide a strip of the same type (same length and color) with a pencil at the position corresponding to the observed position of the beam. The complicated machine is described in great detail: *"Mr. Darwin constructed for us a pendulum, consisting of a bar swinging on knife edges from an axis through its middle point. At either end of the bar were weights, so that by their adjustment very slow or very quick swings could be obtained.... A beam of light from an electric lantern was intercepted by a screen having a thin horizontal slit placed in the slide groove; the selected*

part of the beam reflected from the pendulum mirror was received
on a black screen at some distance from the observers ...". Equally
elaborate was the experimental setup. The experiments were carried out within 1 week, in several sessions. Each session was no
longer that 2 hours to avoid the effect of over-fatigue. All three
individuals observed at the same time. They were separated from
each other by a screen to avoid mutual influence.

The analysis of both series of experiments showed that each of
the three observers looked at the lines with a certain personal bias.
Moreover, each of the individual measurement series showed strong
and persisting serial correlations. Pearson summarized this under
what he called "the personal equation." He concluded: *"So long*
as the variations in the constants of an experimental series can be
shown to be within the errors of random sampling we feel on safe
ground; we know the number of experiments required to obtain a
result with any required degree of accuracy. On the other hand,
when we find significant fluctuations in the personal equation de-
pending on the influence of immediate atmosphere, it becomes all
the more important to show in each individual investigation that
the personal equation itself is insignificant... Note that by "random
sampling" Pearson means that the observations are independent
(or uncorrelated): *A physicist makes twenty or thirty measure-*
ments of a quantity, say by aid of a bright line moving across a
scale. He gives the mean value m of the result and also what he
terms its probable error e. Now the use of this probable error I take
it to be this. If the same experiments were to be repeated by the
same man the same number of times with the mean result $m^{'}$, then
we would expect to find $m^{'} - m$ not a large multiple of the probable
error of the difference $\sqrt{e^2 + e^2} = \sqrt{2}e$." The main message is that,
in spite of laborious experimental precautions, successive observations may be (usually positively) correlated. The reasons that lead
to this dependence are often impossible to discover, not to speak
of describing them quantitatively. Usually, the best we can do is
therefore to model the measurement errors, or deviations from an
overall bias, by a stationary process. Which dependence structure
is suitable has to be analyzed for each data set individually. As
a general tendency one can conclude, however, that cases where
errors exhibit long-range dependence may occur more often than
one would expect a priori. A simple but important consequence
is that the usual $\sigma/\sqrt{n}$ rule for the standard error of the sample
mean needs to be checked each time.

We conclude this section by quoting several other famous scientists who observed the same phenomenon:

Jeffreys (1939, 1948, 1961) discussed Pearson's experiments and the general problem of dependence in his book on probability theory. He used the term *"internal correlation"* and concluded: *"Internal correlation habitually produces such large departures from the usual rule that the standard error of the mean is $n^{-\frac{1}{2}}$ times that of one observation that the rule should never be definitely adopted until it has been checked. In a series of observations made by the same observer, and arranged in order of time, internal correlation is the normal thing, and at the present stage of knowledge hardly needs a significance test any longer. It practically reduces to a problem of estimation. The question of significance arises only when special measures have been taken to eliminate the correlation and we want to know whether they have been successful"*.

Another discussion along this line can be found in Mosteller and Tukey (1977, p.119 ff) in the chapter entitled *"Hunting out the real uncertainty."* In the subsection entitled *"How $\sigma/\sqrt{n}$ can mislead"* they discussed an example by Peirce (1873) that *"exemplifies the history of the personal equation problem of astronomy."* Their comments are analogous to Pearson's conclusions: *"The hope had been that each observer's personal systematic errors could be first stabilized and then adjusted for, thus improving accuracy. Unfortunately, attempts in this direction have failed repeatedly, as these data suggest they might... Wilson and Hilferty (1929) made it clear that Peirce's data illustrate '...that reliance on such formula as $\sigma/\sqrt{n}$ is not scientifically satisfactory in practice, even for estimating unreliability of means' "*. Mosteller and Tukey conclude that *"even in dealing with so simple a statistic as the arithmetic mean, it is often vital to use as direct an assessment of its internal uncertainty as possible. Obtaining a valid measure of uncertainty is not just a matter of looking up a formula."*

1.5.6 Why stationary models? Some "philosophical" remarks

A principal question may be asked at this point. From a scientific point of view, it is not always satisfactory to model an observed phenomenon by a stationary process. After all, there is a reason for everything (is there?). So, should one not try to find the actual cause of (long-range) correlations and construct a corresponding "physical model" ? In particular, this seems most natural in cases where it is obvious that the dependence in the data is mainly caused by some physical mechanism that could be explained, for instance, by a system of differential equations derived from physics (see Section 1.3). A process invented in some general context of theoretical statistics that has nothing to do with the specific problem at hand usually does not give a full physical explanation of why the series behaves like this and not in some other way. A simple model such as (1.9) still might allow a direct physical explanation. Yet, how about long memory? It seems rather strange that events that passed a long time ago should have any influence on events today. A possible view of the matter is the pragmatic approach. In the absence of any useful information that could help us, for example, to give an interval estimate for the expected value or a prediction interval for future observations, a suitable stationary model gives a useful description. If a realistic (and estimable) physical model is not available or cannot be determined with sufficient accuracy, then a simple stationary model is often sufficient, at least for the immediate practical purpose, as it allows us to assess all the essential quantities with reasonable accuracy. This does not mean that one should always stop at this point. On the contrary, it might be, for example, very interesting and practically rewarding to investigate the "personal equation" or to develop a physical theory of the flow of a river in more detail. Stationary processes are a useful tool, in the absence of a feasible physical model. Moreover they can provide valuable help in the search for physical explanations of the observed phenomena. They are, however, not a replacement for "physical" models. On the other hand, it also should be noted that sometimes careful consideration of physical models leads back again to stationary processes.

CHAPTER 2

Stationary processes with long memory

2.1 Introduction

In this chapter we introduce stochastic processes that can be used to model data with the properties discussed in chapter 1. We consider only processes for which the first two moments exist.

The data sets in Chapter 1 had the following common features:

1. Qualitative features of a typical sample path:

a) There are relatively long periods where the observations tend to stay at a high level, and on the other hand, there are long periods with low levels.

b) When one looks only at short time periods, then there seem to be cycles or local trends. However, looking at the whole series, there is no apparent persisting trend or cycle. It rather seems that cycles of (almost) all frequencies occur, superimposed and in random sequence.

c) Overall, the series looks stationary.

In addition to these qualitative features, we also observed the following quantitative properties:

2. The variance of the sample mean seems to decay to zero at a slower rate than n^{-1}. In good approximation, the rate is proportional to $n^{-\alpha}$ for some $0 < \alpha < 1$.

3. The sample correlations $\hat{\rho}(k) = \hat{\gamma}(k)/\hat{\gamma}(0)$ decay to zero at a rate that is in good approximation proportional to $k^{-\alpha}$ for some $0 < \alpha < 1$.

4. Near the origin, the logarithm of the periodogram $I(\lambda)$ plotted against the logarithm of the frequency appears to be randomly scattered around a straight line with negative slope.

Point 1c) implies that at least to a first approximation, it is reasonable to assume stationarity. Let us therefore assume that the data are a sample path of a stationary process X_t. We can reformulate properties 2 to 4 as mathematical conditions on the stationary process:

2. The variance of the sample mean $\text{var}(\bar{X})$ is asymptotically equal to a constant c_{var} times $n^{-\alpha}$ for some $0 < \alpha < 1$.

3. The correlations $\rho(k)$ are asymptotically equal to a constant c_ρ times $k^{-\alpha}$ for some $0 < \alpha < 1$.

4. The spectral density $f(\lambda)$ has a pole at zero that is equal to a constant c_f times $\lambda^{-\beta}$ for some $0 < \beta < 1$.

A slight generalization of these conditions may be obtained by replacing the proportionality constants $c_{var}, c_\rho,$ and c_f by so-called slowly varying functions, i.e., functions such that for any $t \in R$, $L(tx)/L(x) \to 1$ as $x \to \infty$ (or $x \to 0$, respectively). For most practical purposes, this generalization is not needed, however. Throughout the book, we therefore use conditions 2 to 4, though most of the results also hold for the more general case. Thus, we will use the following definition of a stationary process with long memory or long-range dependence:

Definition 2.1 *Let X_t be a stationary process for which the following holds. There exists a real number $\alpha \in (0,1)$ and a constant $c_\rho > 0$ such that*

$$\lim_{k \to \infty} \rho(k)/[c_\rho k^{-\alpha}] = 1. \qquad (2.1)$$

Then X_t is called a stationary process with long memory or long-range dependence or strong dependence, or a stationary process with slowly decaying or long-range correlations.

For reasons, we will see later, the parameter $H = 1 - \alpha/2$ will also be used instead of α. In terms of this parameter, long memory occurs for $\frac{1}{2} < H < 1$. Knowing the covariances (or correlations and variance) is equivalent to knowing the spectral density f. Therefore, long-range dependence can also be defined by imposing a condition on the spectral density:

Definition 2.2 *Let X_t be a stationary process for which the following holds: There exists a real number $\beta \in (0,1)$ and a constant $c_f > 0$ such that*

$$\lim_{\lambda \to 0} f(\lambda)/[c_f |\lambda|^{-\beta}] = 1. \qquad (2.2)$$

Then X_t is called a stationary process with long memory or long-range dependence or strong dependence.

These two definitions are equivalent in the following sense (Zygmund 1959, Chapter V.2):

Theorem 2.1 *(i) Suppose (2.1) holds with $0 < \alpha = 2 - 2H < 1$. Then the spectral density f exists and*

$$\lim_{\lambda \to 0} f(x)/[c_f(H)|\lambda|^{1-2H}] = 1, \qquad (2.3)$$

where

$$c_f = \sigma^2 \pi^{-1} c_\rho \Gamma(2H - 1) \sin(\pi - \pi H) \qquad (2.4)$$

and $\sigma^2 = \mathrm{var}(X_t)$.

(ii) Suppose (2.2) holds with $0 < \beta = 2H - 1 < 1$. Then

$$\lim_{k \to \infty} \rho(k)/[c_\rho k^{2H-2}] = 1, \qquad (2.5)$$

where

$$c_\rho = \frac{c_\gamma}{\sigma^2} \qquad (2.6)$$

and

$$c_\gamma = 2c_f \Gamma(2 - 2H) \sin(\pi H - \frac{1}{2}\pi). \qquad (2.7)$$

It is important to note that the definition of long-range dependence by (2.1) [or (2.2)] is an asymptotic definition. It only tells us something about the ultimate behavior of the correlations as the lag tends to infinity. In this generality, it does not specify the correlations for any fixed finite lag. Moreover, it determines only the rate of convergence, not the absolute size. Each individual correlation can be arbitrarily small. Only the decay of the correlations is slow. This makes the detection of slowly decaying correlations more difficult. A standard method for checking for non-zero correlations is, for example, to look at the $\pm 2/\sqrt{n}$-confidence band in the plot of $\hat{\rho}(k)$ against k (see, e.g., Priestley 1981, p. 340). A sample correlation $\hat{\rho}(k)$ is considered significant if $|\hat{\rho}(k)| > 2/\sqrt{n}$. For slowly decaying correlations, it can easily happen that all $\hat{\rho}(k)$s are within this interval. An example illustrates this difficulty: Suppose that the correlations are given by $\rho(k) = c|k|^{2H-2}$ with $c = 0.1$ and $H = 0.9$. Figure 2.1a displays $\rho(k)$ together with the $\pm 2/\sqrt{n}$-bands for $n = 100, 200, 400$, and 1000. We see that even if our estimated correlations were exactly equal to the correct correlations, we would not consider them significant unless n is at least 400. Nevertheless, the effect on statistical inference is by far not negligible even for small sample sizes. The ratios $[\mathrm{var}(\bar{X})]^{1/2}/[\sigma/\sqrt{n}\,]$ displayed in Figure 2.1b show that confidence intervals for $\mu = E(X_t)$ based on the assumption of independence are too small by a factor

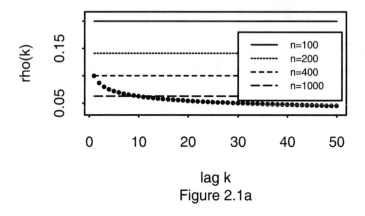

lag k

Figure 2.1a

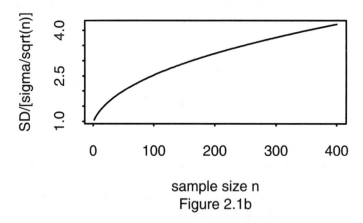

sample size n

Figure 2.1b

Figure 2.1. *Autocorrelations* $0.1 \cdot k^{-0.2}$ *and* $2n^{-\frac{1}{2}}$ *confidence bands (Figure 2.1a); standard deviation of the sample mean compared to* $\sigma n^{-\frac{1}{2}}$ *(Figure 2.1b).*

of about 2 for $n = 100$ and 4 for $n = 400$. This happens notwith-standing the fact that each individual correlation is very small. It is the slow decay, and thus the joint effect of all correlations to-gether, that increases the variance of the sample mean. Therefore, to detect such dependence, it is better to consider all correlations simultaneously, instead of considering individual correlations sep-arately. Considering the speed at which the correlations converge to zero gives us a better idea about the order of magnitude of $\mathrm{var}(\bar{X}_n)$.

Let us now see if long-memory processes as defined above fulfill our postulated properties 1 to 4. It is clear that the definitions directly imply 3 and 4. How about 1a to 1c and 2 ? The following result shows that condition 2 follows from definition 2.2 (or 2.3) (see, e.g., Beran 1989a):

Theorem 2.2 *Let X_t be a stationary process with long-range de-pendence. Then*

$$\lim_{n\to\infty} \mathrm{var}(\sum_{i=1}^{n} X_i)/[c_\gamma n^{2H}] = \frac{1}{H(2H-1)}. \tag{2.8}$$

It remains to be shown that the qualitative behavior of typical sample paths corresponds to 1a to 1c. Figures 2.2 and 2.3 show typical sample paths of two different processes, with no memory and long memory of varying degrees. The processes displayed are fractional Gaussian noise and fractional $\mathrm{ARIMA}(0, H - \frac{1}{2}, 0)$ with $H = 0.5, 0.7, 0.9$ (for the definitions see Sections 2.4 and 2.5). The plots exhibit the behavior we wanted to achieve. The characteristic features 1a and 1b are the more prominent the higher the value of H is, i.e., the stronger the long-range dependence is. In the case of $H = \frac{1}{2}$ we have, in both cases (Figures 2.2a and 2.3a), independent identically distributed observations.

Among the many possible models for which (2.1) holds, there are two classes of models that are of special interest, because they arise in a natural way from limit theorems and classic models, respectively: (1) stationary increments of self-similar processes, in particular fractional Gaussian noise, and (2) fractional ARIMA processes. We will now discuss these two classes.

2.2 Self-similar processes

Self-similar processes were introduced by Kolmogorov (1941) in a theoretical context. Statisticians do not seem to have been aware

Figure 2.2a

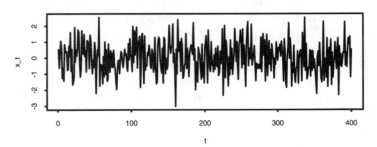

Figure 2.2b

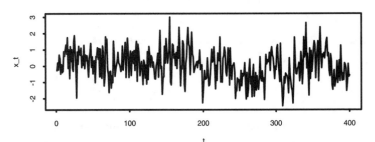

Figure 2.2c

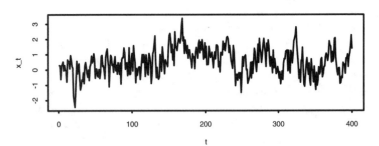

Figure 2.2. *Simulated series of fractional Gaussian noise with $H = 0.5$ (Figure 2.2a), $H = 0.7$ (Figure 2.2b), and $H = 0.9$ (Figure 2.2c).*

Figure 2.3a

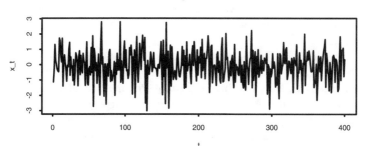

Figure 2.3b

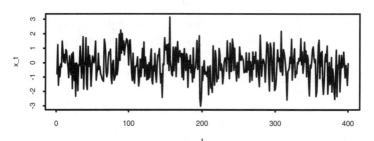

Figure 2.3c

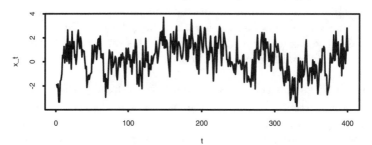

Figure 2.3. *Simulated series of a fractional ARIMA(0,d,0) process with* $H = 0.5$ *(Figure 2.3a),* $H = 0.7$ *(Figure 2.3b), and* $H = 0.9$ *(Figure 2.3c).*

of the existence or statistical relevance of such processes, until
Mandelbrot and his co-workers (see, e.g., Mandelbrot and van
Ness 1968, Mandelbrot and Wallis 1969a,b,c) introduced them into
statistics.

The basic idea of self-similarity is much older. Mandelbrot (1977,
1983a) refered, for example, to Leonardo da Vinci's drawings of
turbulent flows that exhibit coexistent "eddies" of all sizes and
thus self-similarity. A geometric shape is called self-similar in a
deterministic way if the same geometric structures are observed,
independently of the distance from which one looks at the shape
(for an exact mathematical definition see Mandelbrot 1977, 1983a).
In the context of stochastic processes, self-similarity is defined in
terms of the distribution of the process:

Definition 2.3 *Let Y_t be a stochastic process with continuous time
parameter t. Y_t is called self-similar with self-similarity parameter
H, if for any positive stretching factor c, the rescaled process with
time scale ct, $c^{-H}Y_{ct}$, is equal in distribution to the original process
Y_t.*

This means that, for any sequence of time points $t_1, ..., t_k$, and
any positive constant c, $c^{-H}(Y_{ct_1}, Y_{ct_2}, ..., Y_{ct_k})$ has the same distribution as $(Y_{t_1}, Y_{t_2}, ..., Y_{t_k})$. Thus, typical sample paths of a self-
similar process look qualitatively the same, irrespective of the distance from which we look at them. In contrast to deterministic
self-similarity, it does not mean that the same picture repeats itself exactly as we go closer. It is rather the general impression that
remains the same. Figure 2.4 illustrates this on a typical sample
path of fractional Brownian motion. A definition of this process is
given below.

What is the motivation behind stochastic self-similarity ? Apart
from the aesthetic appeal and mathematical elegance, there is
a more profound justification of self-similarity. Lamperti (1962)
showed that self-similarity (he used the term "semi-stability")
arises in a natural way from limit theorems for sums of random variables (see also Vervaat 1987). The following definition is needed.

Definition 2.4 *If for any $k \geq 1$ and any k time points $t_1, ..., t_k$,
the distribution of $(Y_{t_1+c} - Y_{t_1+c-1}, ..., Y_{t_k+c} - Y_{t_k+c-1})$ does not
depend on $c \in R$, then we say that Y_t has stationary increments.*

The theorem can now be stated as follows:

Theorem 2.3 *(i) Suppose that Y_t is a stochastic process such
that $Y_1 \neq 0$ with positive probability and Y_t is the limit in distri-*

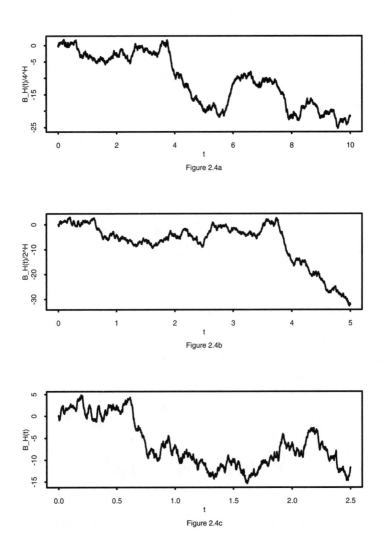

Figure 2.4. *Fractional Brownian motion with* $H = 0.7, 0 < t < 10$ *(Figure 2.4a); normalized sample path as in Figure 2.4a for* $0 < t < 5$ *(Figure 2.4b); normalized sample path as in Figure 2.4a for* $0 < t < 2.5$ *(Figure 2.4c).*

bution of the sequence of normalized partial sums

$$a_n^{-1} S_{nt} = a_n^{-1} \sum_{i=1}^{[nt]} X_i, \ n = 1, 2, \dots$$

Here $[nt]$ denotes the integer part of nt, $X_1, X_2, \dots$ is a stationary sequence of random variables, and $a_1, a_2, \dots$ is a sequence of positive normalizing constants such that $\log a_n \to \infty$. Then there exists an $H > 0$ such that for any $u > 0$,

$$\lim_{n \to \infty} \frac{a_{nu}}{a_n} = u^H,$$

and Y_t is self-similar with self-similarity parameter H, and has stationary increments.

(ii) All self-similar processes with stationary increments and $H > 0$ can be obtained by partial sums as given in (i).

Part (i) of this theorem essentially says that whenever a process is the limit of normalized partial sums of random variables, it is necessarily self-similar. Thus one can say that, as a result of Lamperti's limit theorem, the role of self-similar processes among stochastic processes is analogous to the central role of stable distributions among distributions.

Self-similarity can be defined analogously for stochastic processes Y_t with a multidimensional "time"-parameter t. Multiplication by a constant is then replaced by affine transformations or, more generally, a group of transformations. Such processes play an important role in the study of critical phenomena in physics, in particular in the so-called "renormalization theory" (see Section 1.3.3). It is beyond the scope of this book to discuss this theory in detail. For probabilistic results in this context we refer, for example, to Vervaat (1987), Laha and Rohatgi (1982), and Hudson and Mason (1982).

2.3 Stationary increments of self-similar processes

Suppose that Y_t is a self-similar process with self-similarity parameter H. The property

$$Y_t =_d t^H Y_1 \ (\text{ for } t > 0), \tag{2.9}$$

where $=_d$ is equality in distribution, implies the following limiting behavior of Y_t as t tends to infinity (see Vervaat 1987):

1. If $H < 0$, then $Y_t \to_d 0$ (where $\to_d$ is convergence in distribution).

2. If $H = 0$, then $Y_t =_d Y_1$.

3. If $H > 0$ and $Y_t \neq 0$, then $|Y_t| \to_d \infty$.

Analogously, for t converging to zero, we have:

1. If $H < 0$ and $Y_t \neq 0$, then $|Y_t| \to_d \infty$.

2. If $H = 0$, then $Y_t =_d Y_1$.

3. If $H > 0$, then $Y_t \to_d 0$.

If we exclude the trivial case $Y_t \equiv 0$, then these properties imply that Y_t is not stationary unless $H = 0$. The exception $H = 0$ is not interesting, as it implies that for all $t > 0$, Y_t is equal to Y_1 with probability 1. For the purpose of modelling data that look stationary, we need only to consider self-similar processes with stationary increments. The range of H can then be restricted to $H > 0$. The reason is that if the increments of a self-similar process are stationary, then the process is mathematically pathological for negative values of H. More specifically, for $H < 0$, Y_t is not a measurable process (Vervaat 1985, 1987). The only exception is the trivial case where $Y_t = Y_1 = 0$ with probability 1. A proof and more detailed mathematical explanations are given, for example, in Vervaat (1985, 1987). Thus, in the following, we consider positive values of H only, in particular, $Y_o = 0$ with probability 1.

The form of the covariance function $\gamma_y(t, s) = \mathrm{cov}(Y_t, Y_s)$ of a self-similar process Y_t with stationary increments follows from these two properties. To simplify notation, assume $E(Y_t) = 0$. Let $s < t$ and denote by $\sigma^2 = E[(Y_t - Y_{t-1})^2] = E[Y_1^2]$ the variance of the increment process $X_t = Y_t - Y_{t-1}$. Then

$$E[(Y_t - Y_s)^2] = E[(Y_{t-s} - Y_o)^2] = \sigma^2(t - s)^{2H}.$$

On the other hand,

$$E[(Y_t - Y_s)^2] = E[Y_t^2] + E[Y_s^2] - 2E[Y_t Y_s] = \sigma^2 t^{2H} + \sigma^2 s^{2H} - 2\gamma_y(t, s).$$

Hence,

$$\gamma_y(t, s) = \frac{1}{2}\sigma^2[t^{2H} - (t - s)^{2H} + s^{2H}]. \qquad (2.10)$$

Similarly, the covariances of the increment sequence $X_i = Y_i - Y_{i-1}$ $(i = 1, 2, 3, ...)$ are obtained. The covariance between X_i and X_{i+k} $(k > 0)$ is equal to

$$\gamma(k) = \mathrm{cov}(X_i, X_{i+k}) = \mathrm{cov}(X_1, X_{k+1})$$

$$= \frac{1}{2}E[(\sum_{j=1}^{k+1} X_j)^2 + (\sum_{j=2}^{k} X_j)^2 - (\sum_{j=1}^{k} X_j)^2 - (\sum_{j=2}^{k+1} X_j)^2]$$

$$= \frac{1}{2}\{E[(Y_{k+1} - Y_o)^2] + E[(Y_{k-1} - Y_o)^2]$$
$$- E[(Y_k - Y_o)^2] - E[(Y_k - Y_o)^2]\}.$$

Using self-similarity, we obtain the formula

$$\gamma(k) = \frac{1}{2}\sigma^2[(k+1)^{2H} - 2k^{2H} + (k-1)^{2H}] \tag{2.11}$$

for $k \geq 0$ and $\gamma(k) = \gamma(-k)$ for $k < 0$. The correlations are given by

$$\rho(k) = \frac{1}{2}[(k+1)^{2H} - 2k^{2H} + (k-1)^{2H}] \tag{2.12}$$

for $k \geq 0$ and $\rho(k) = \rho(-k)$ for $k < 0$.

The asymptotic behavior of $\rho(k)$ follows by Taylor expansion: First note that $\rho(k) = \frac{1}{2}k^{2H}g(k^{-1})$ where $g(x) = (1 + x)^{2H} - 2 + (1 - x)^{2H}$. If $0 < H < 1$ and $H \neq 1/2$, then the first non-zero term in the Taylor expansion of $g(x)$, expanded at the origin, is equal to $2H(2H - 1)x^2$. Therefore, as k tends to infinity, $\rho(k)$ is equivalent to $H(2H - 1)k^{2H-2}$, i.e.,

$$\rho(k)/[H(2H - 1)k^{2H-2}] \to 1 \tag{2.13}$$

as $k \to \infty$. For $1/2 < H < 1$, this means that the correlations decay to zero so slowly that

$$\sum_{k=-\infty}^{\infty} \rho(k) = \infty \tag{2.14}$$

The process X_i $(i = 1, 2, ...)$ has long memory. For $H = 1/2$, all correlations at non-zero lags are zero, i.e., the observations X_i are uncorrelated. For $0 < H < 1/2$, the correlations are summable. In fact a more specific equation holds, namely,

$$\sum_{k=-\infty}^{\infty} \rho(k) = 0. \tag{2.15}$$

In practice, this case is rarely encountered (though it may occur after overdifferencing), mainly because condition (2.15) is very unstable. A simple example illustrates this: Suppose that (2.15) holds for Y_t. Let δ_t be a process that is independent of Y_t, with an arbitrarily small variance $\text{var}(\delta_t)$ and with autocorrelations that sum up to a positive (finite or infinite) constant c. Assume that instead of Y_t, we observe the slightly disturbed process $Y_t + \delta_t$. Then the sum of the correlations of the observed process $Y_t + \delta_t$ is equal to $c \neq 0$. Thus, an arbitrarily small disturbance destroys property (2.15).

So far, we considered the cases $0 < H < \frac{1}{2}, H = \frac{1}{2}$ and $\frac{1}{2} < H < 1$. What happens for $H \geq 1$? For $H = 1$, (2.9) implies $\rho(k) \equiv 1$. All correlations are equal to 1, no matter how far apart in time the observations are. This case is hardly of any practical importance. For $H > 1$, $g(k^{-1})$ diverges to infinity. This contradicts the fact that $\rho(k)$ must be between -1 and 1. We can conclude that if covariances exist and $\lim_{k \to \infty} \rho(k) = 0$, then

$$0 < H < 1. \tag{2.16}$$

For $\frac{1}{2} < H < 1$ the process has long-range dependence, for $H = \frac{1}{2}$ the observations are uncorrelated, and for $0 < H < \frac{1}{2}$ the process has short-range dependence and the correlations sum up to zero.

It should be stressed that here we consider only processes with finite second moments. There are well-defined self-similar processes with stationary increments, infinite second moments, and $H \geq 1$ (see the references in Section 11.1). Also, one should add that, if the second moments do not exist, then $H \neq \frac{1}{2}$ does not necessarily imply that the increments X_i are dependent. For example there are so-called α-stable processes with $0 < \alpha < 2$ that are self-similar with self-similarity parameters $H = 1/\alpha$ and have independent increments, although $H = 1/\alpha > \frac{1}{2}$ (see, e.g., Samorodnitsky and Taqqu 1993). Unless specified otherwise, we will always assume that the second moments are finite, $\lim_{k \to \infty} \rho(k) = 0$, and hence $0 < H < 1$. Under these assumptions, the spectral density of the increment process X_i can be derived from (2.7) (Sinai 1976):

Proposition 2.1 *The spectral density of X_i is given by*

$$f(\lambda) = 2c_f(1 - \cos\lambda) \sum_{j=-\infty}^{\infty} |2\pi j + \lambda|^{-2H-1}, \ \lambda \in [-\pi, \pi] \tag{2.17}$$

with $c_f = c_f(H, \sigma^2) = \sigma^2(2\pi)^{-1}\sin(\pi H)\Gamma(2H + 1)$ and $\sigma^2 = \mathrm{var}(X_i)$.

The behavior of f near the origin follows by Taylor expansion at zero:

Corollary 2.1 *Under the above assumptions*

$$f(\lambda) = c_f|\lambda|^{1-2H} + O(|\lambda|^{\min(3-2H,2)}). \tag{2.18}$$

The approximation of f by $c_f|\lambda|^{1-2H}$ is in fact very good even for relatively large frequencies. This is illustrated in Figure 2.5.

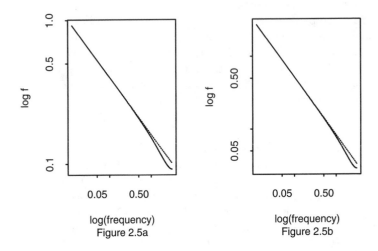

Figure 2.5. *Spectral density of fractional Gaussian noise with* $H = 0.7$ *(Figure 2.5a) and* $H = 0.9$ *(Figure 2.5b) and approximation by straight lines.*

There, $\log f(\lambda)$ and $y = \log c_f + (1 - 2H) \log \lambda$, respectively, are plotted against $\log \lambda$. The logarithm of f deviates only very little from the corresponding straight line.

We conclude this Section by noting an appealing property of stationary increments of self-similar processes: The sample mean can be written as

$$\bar{X} = n^{-1} \sum_{i=1}^{n} X_i = n^{-1}(Y_n - Y_o) =_d n^{-1} n^H (Y_1 - Y_o).$$

Therefore, instead of the asymptotic equality (2.8), we obtain for each sample size the exact equality

$$\text{var}(\bar{X}) = \sigma^2 n^{2H-2}. \tag{2.19}$$

For $H = \frac{1}{2}$, this is the classic result $\text{var}(\bar{X}) = \sigma^2 n^{-1}$. Moreover, if X_i is a Gaussian process with mean μ and variance σ^2, then $n^{1-H}(\bar{X} - \mu)/\sigma$ is a standard normal random variable. This can be used to calculate tests and confidence intervals for μ.

2.4 Fractional Brownian motion and Gaussian noise

Suppose that Y_t is a self-similar process with stationary increments. The expected value of the increment process $X_i = Y_i - Y_{i-1}$ ($i = 1, 2, ...$) is zero. The covariances of X_i are given by equation (2.11). In particular, suppose that X_t is a Gaussian process. Then, the distribution of the process is fully specified by the mean and covariances. Therefore, for each value of $H \in (0, 1)$, there is exactly one Gaussian process X_i that is the stationary increment of a self-similar process Y_t. This process is called *fractional Gaussian noise*. The corresponding self-similar process Y_t is called *fractional Brownian motion*. We will denote it by $B_H(t)$.

Let us first consider the case $H = \frac{1}{2}$. For $H = \frac{1}{2}$, $X_1, X_2, ...$ are independent normal variables. The corresponding self-similar process $B_{\frac{1}{2}}(t)$ turns out to be ordinary Brownian motion, which we will denote by $B(t)$. This can be seen by deriving self-similarity with $H = \frac{1}{2}$ from the following standard definition of Brownian motion:

Definition 2.5 *Let $B(t)$ be a stochastic process with continuous sample paths and such that*

(i) $B(t)$ is Gaussian,

(ii) $B(0) = 0$ almost surely (a.s.)

(iii) $B(t)$ has independent increments,

(iv) $E[B(t) - B(s)] = 0$

(v) $\mathrm{var}[B(t) - B(s)] = \sigma^2 |t - s|$.

Then $B(t)$ is called Brownian motion.

Self-similarity with $H = \frac{1}{2}$ follows directly from this definition: First note that, because $B(t)$ is Gaussian, it is sufficient to look at the expected value and the covariances of $B(t)$. From (ii) and (iv) we have

$$E[B(ct)] = E[B(ct) - B(0)] = 0 = c^{\frac{1}{2}} E[B(t)].$$

Consider now the covariance $\mathrm{cov}(B(t), B(s))$ for $t > s$. Because $B(t) - B(s)$ is independent of $B(s) - B(0) = B(s)$, we can write

$$\mathrm{cov}(B(t), B(s)) = \mathrm{var}(B(s) - B(0))$$

$$= \sigma^2 s = \sigma^2 \min(t, s).$$

Therefore, for any $c > 0$

$$\mathrm{cov}(B(ct), B(cs)) = c\sigma^2 \min(t, s)$$

$$= \text{cov}(c^{1/2}B(t), c^{1/2}B(s)).$$

Hence, $B(t)$ is self-similar with self-similarity parameter $H = \frac{1}{2}$.

Fractional Brownian motion with H assuming any value in the interval $(0, 1)$ is defined by the covariance function (2.11). Alternatively, it can be defined as a weighted average of ordinary Brownian motion over the infinite past. A mathematically stringent definition along this line can be given in terms of a stochastic integral:

Definition 2.6 *Let $s > 0$ be a positive scaling constant, and define the weight function w_H by*

$$w_H(t, u) = 0 \text{ for } t \leq u,$$

$$w_H(t, u) = (t - u)^{H - \frac{1}{2}} \text{ for } 0 \leq u < t,$$

and

$$w_H(t, u) = (t - u)^{H - \frac{1}{2}} - (-u)^{H - \frac{1}{2}} \text{ for } u < 0.$$

Also, let $B(t)$ be standardized Brownian motion, i.e., $B(t)$ is as in Definition 2.5 with $\sigma^2 = 1$. For $0 < H < 1$, let $B_H(t)$ be defined by the stochastic integral

$$B_H(t) = s \int w_H(t, u) dB(u), \tag{2.20}$$

where the convergence of the integral is to be understood in the L^2-norm with respect to the Lebesgue measure on the real numbers. Then $B_H(t)$ is called fractional Brownian motion with self-similarity parameter H.

For the exact mathematical definition of such stochastic integrals we refer the interested reader to Ash and Gardner (1975). A brief intuitive explanation can be given as follows: The function $w_H(t, .)$ is first approximated by a sequence of step functions:

$$w_{H,m}(t, u) = \sum_{j=1}^{m} c_j 1\{u_{j-1} < u < u_j\},$$

where $-\infty < u_o < u_1 < ... < u_m < \infty$. We choose a sequence of step functions $w_{H,m}$ such that

$$\int [w_H(t, u) - w_{H,m}(t, u)]^2 du \rightarrow 0$$

as $m \rightarrow \infty$. For $w_{H,m}(t, u)$, the stochastic integral with respect to Brownian motion is defined to be equal to

$$S_m = \int w_{H,m}(t, u) dB(u) = \sum_{j=1}^{m} c_j [B(u_j) - B(u_{j-1})].$$

The integral (2.19) is then defined by

$$\int w_H(t,u)dB(u) = \lim_{m\to\infty} \int w_{H,m}(t,u)dB(u) = \lim_{m\to\infty} S_m.$$

The limit is to be understood in the sense of L^2-convergence. This means that for a single path of $B(u)$, the limit does not need to exist. However, S_m converges in mean square, i.e., there exists a random variable S such that

$$\lim_{m\to\infty} E[(S_m - S)^2] = 0.$$

Self-similarity of the weighted integral (2.19), with self-similarity parameter H, follows directly from self-similarity of $B(t)$ and the definition of $w_H(t,u)$: First note that

$$w(ct, u) = c^{H-\frac{1}{2}} w(t, uc^{-1}).$$

Therefore,

$$B_H(ct) = \int w_H(ct,u)dB(u) = c^{H-\frac{1}{2}} \int w_H(t, uc^{-1})dB(u).$$

Using the substitution $v = uc^{-1}$, we obtain

$$c^{H-\frac{1}{2}} \int w_H(t,v)dB(cv).$$

By self-similarity of $B(t)$, this is equal in distribution to

$$c^{H-\frac{1}{2}} c^{\frac{1}{2}} \int w_H(t,v)dB(v) = c^H B_H(t).$$

Thus, $B_H(t)$ defined by (2.19) is self-similar with self-similarity parameter H.

It is informative to take a closer look at the weight function $w_H(t,u)$: Figure 2.6 shows $w_H(t,u)$ as a function of u for $t = 1$ and several values of H. For $H = \frac{1}{2}$,

$$w_H(t,u) = 1, \text{ for } 0 < u \leq t$$

and

$$w_H(t,u) = 0, \text{ for } u \leq 0 \text{ and for } u > t.$$

This means that $B_H(t) = B(t)$ and the increments are independent identically distributed normal random variables.

If H is in the intervals $(0, 1/2)$ or $(1/2, 1)$, then the weight function is proportional to $|u|^{H-3/2}$ as u tends to $-\infty$. For $H > 1/2$,

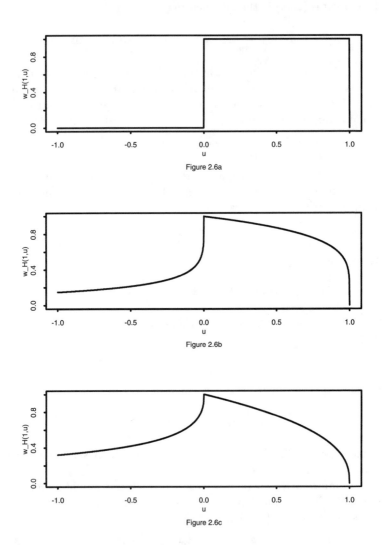

Figure 2.6. *Weight function* $w_H(t, u)$ *for fractional Brownian motion with* $H = 0.5$ *(Figure 2.6a),* $H = 0.7$ *(Figure 2.6b),* $H = 0.9$ *(Figure 2.6c), and* $t = 1$.

$|u|^{H-3/2}$ tends to zero so slowly that

$$\int_{-\infty}^{t} w_H(t,u)du = \infty \qquad (2.21)$$

for all $t \in R$. For $H < 1/2$, $u^{H-3/2}$ dies off very quickly and

$$\int_{-\infty}^{t} w_H(t,u)du = 0. \qquad (2.22)$$

These properties are reflected in the corresponding properties of the correlations $\rho(k)$ for the differenced process, (2.14) and (2.15).

2.5 Fractional ARIMA models

ARIMA models were introduced by Box and Jenkins (1970). Because of their simplicity and flexibility, they became very popular in applied time series analysis. The theory of statistical inference for these processes is well developed. A detailed account of statistical methods and references can be found in, for example, Box and Jenkins (1970), Priestley (1981), Abraham and Ledolter (1983), and Brockwell and Davis (1987). Fractional ARIMA models are a natural extension of the classic ARIMA models.

Let us first recall the definition of ARMA and ARIMA processes. To simplify notation, we assume $\mu = E(X_t) = 0$. Otherwise, X_t must be replaced by $X_t - \mu$ in all formulas. In the following, B will denote the backshift operator B defined by $BX_t = X_{t-1}$, $B^2X_t = X_{t-1}, ...$ In particular, differences can be expressed in terms of the backshift operator as $X_t - X_{t-1} = (1 - B)X_t$, $(X_t - X_{t-1}) - (X_{t-1} - X_{t-2}) = (1-B)^2X_t, ...$ Let p and q be integers. Define the polynomials

$$\phi(x) = 1 - \sum_{j=1}^{p} \phi_j x^j$$

and

$$\psi(x) = 1 + \sum_{j=1}^{q} \psi_j x^j.$$

Assume that all solutions of $\phi(x_o) = 0$ and $\psi(x) = 0$ are outside the unit circle. Furthermore, let ϵ_t $(t = 1, 2, ...)$ be iid normal variables with zero expectation and variance σ_ϵ^2. An ARMA(p,q) model is defined to be the stationary solution of

$$\phi(B)X_t = \psi(B)\epsilon_t. \qquad (2.23)$$

If instead (2.23) holds for the dth difference $(1 - B)^d X_t$, then X_t is called an ARIMA(p, d, q) process. The corresponding equation is

$$\phi(B)(1 - B)^d X_t = \psi(B)\epsilon_t. \qquad (2.24)$$

Note that an ARMA(p, q) process is also an ARIMA$(p, 0, q)$ process. If d is larger than or equal to 1, then the original series X_t is not stationary. To obtain a stationary process, X_t must be differenced d times.

Equation (2.24) can be generalized in a natural way by allowing d to assume any real value. First note that if d is an integer $(d \geq 0)$, then $(1 - B)^d$ can also be written as

$$(1 - B)^d = \sum_{k=0}^{d} \binom{d}{k} (-1)^k B^k,$$

with the binomial coefficients

$$\binom{d}{k} = \frac{d!}{k!(d - k)!} = \frac{\Gamma(d + 1)}{\Gamma(k + 1)\Gamma(d - k + 1)}.$$

Here, $\Gamma(x)$ denotes the gamma function. As the gamma function is also defined for all real numbers, the above definition of the binomial coefficient can be extended to all real numbers d. Note that for negative integers the gamma function has poles so that the binomial coefficient is zero if $k > d$ and d is an integer. Formally, we can define $(1 - B)^d$ for any real number d by

$$(1 - B)^d = \sum_{k=0}^{\infty} \binom{d}{k} (-1)^k B^k. \qquad (2.25)$$

For all positive integers, only the first $d + 1$ terms are non-zero and we obtain the original definition of the dth difference operator $(1 - B)^d$. For non-integer values of d, the summation in (2.25) is genuinely over an infinite number of indices. Definition (2.24) can now be extended to non-integer values of d in the following way:

Definition 2.7 *Let X_t be a stationary process such that*

$$\phi(B)(1 - B)^d X_t = \psi(B)\epsilon_t \qquad (2.26)$$

for some $-\frac{1}{2} < d < \frac{1}{2}$. *Then X_t is called a fractional ARIMA(p, d, q) process.*

This definition was proposed by Granger and Joyeux (1980) and Hosking (1981). The range that is interesting in the context of long-memory processes is $0 \leq d < \frac{1}{2}$. The upper bound $d < \frac{1}{2}$

is needed, because for $d \geq \frac{1}{2}$, the process is not stationary, at least not in the usual sense. In particular, the usual definition of the spectral density of X_t would lead to a nonintegrable function [see (2.27) below]. For the range $\frac{1}{2} \leq d \leq 1$, one can still define a "spectral density," by using a more general definition. It is however not integrable (see the remarks in Section 1.3.5). Also note that, the case $d > \frac{1}{2}$ can be reduced to the case $-\frac{1}{2} < d \leq \frac{1}{2}$ by taking appropriate differences. For instance, if (2.26) holds with $d = 1.3$, then the differenced process $X_t - X_{t-1}$ is the stationary solution of (2.26) with $d = 0.3$.

Equation (2.26) can be interpreted in several ways. For instance, it can be written as

$$(1 - B)^d X_t = \tilde{X}_t, \tag{2.27}$$

where $\tilde{X}_t$ is an ARMA process defined by

$$\tilde{X}_t = \phi^{-1}(B)\psi(B)\epsilon_t \tag{2.28}$$

This means that, after passing X_t through the fractional difference operator (or infinite linear filter) $(1 - B)^d$, we obtain an ARMA process. On the other hand, we can write

$$X_t = \phi(B)^{-1}\psi(B)X_t^*, \tag{2.29}$$

where X_t^* is a fractional ARIMA$(0, d, 0)$ process defined by

$$X_t^* = (1 - B)^{-d}\epsilon_t \tag{2.30}$$

That is, X_t is obtained by passing a fractional ARIMA$(0, d, 0)$ process through an ARMA filter. Figures 2.7a to c show sample paths of several fractional ARIMA processes. We can see that many different types of behavior can be obtained. The parameter d determines the long-term behavior, whereas p, q, and the corresponding parameters in $\phi(B)$ and $\psi(B)$ allow for more flexible modelling of short-range properties.

The spectral density of a fractional ARIMA process follows directly from (2.26). Denote by

$$f_{ARMA}(\lambda) = \frac{\sigma_\epsilon^2}{2\pi} \frac{|\psi(e^{i\lambda})|^2}{|\phi(e^{i\lambda})|^2}$$

the spectral density of the ARMA process $\tilde{X}_t$. Recall that if X_t is obtained from a process Y_t with spectral density f_Y by applying the linear filter $\sum a(s)Y_{t-s}$, then the spectral density of X_t is equal to $|A(\lambda)|^2 f_Y(\lambda)$, where $A(\lambda) = \sum a(s)e^{is\lambda}$ (see, e.g., Priestley 1981,

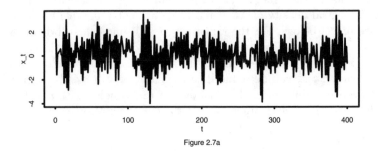

Figure 2.7a

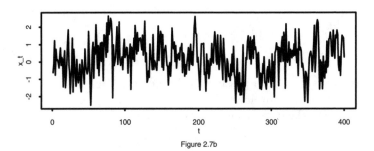

Figure 2.7b

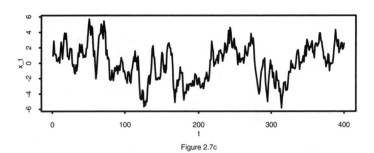

Figure 2.7c

Figure 2.7. *Simulated series of a fractional ARIMA(1,0.3,0) process with AR parameter −0.7 (Figure 2.7a), a fractional ARIMA(0,0.3,0) process (Figure 2.7b), and a fractional ARIMA(1,0.3,0) process with AR parameter 0.7 (Figure 2.7c).*

p. 266). From (2.26) we then obtain the spectral density of X_t :

$$f(\lambda) = |1 - e^{i\lambda}|^{-2d} f_{ARMA}(\lambda). \qquad (2.31)$$

Note that $|1 - e^{i\lambda}| = 2 \sin \frac{1}{2}\lambda$. Because $\lim_{\lambda \to 0} \lambda^{-1}(2 \sin \frac{1}{2}\lambda) = 1$, the behavior of the spectral density at the origin is given by

$$f(\lambda) \sim \frac{\sigma_\epsilon^2}{2\pi} \frac{|\psi(1)|^2}{|\phi(1)|^2} |\lambda|^{-2d} = f_{ARMA}(0)|\lambda|^{-2d}. \qquad (2.32)$$

Thus, for $d > 0$, the spectral density has a pole at zero. Comparing this with our notation in the previous sections we see that

$$d = H - \frac{1}{2}. \qquad (2.33)$$

For $d = 0$, X_t is an ordinary ARMA$(p, 0, q)$ process with bounded spectral density. Long-range dependence occurs for

$$0 < d < \frac{1}{2}. \qquad (2.34)$$

In order to transform X_t into a process with a bounded spectral density, one has to apply the linear filter $(1 - B)^d$. For $-\frac{1}{2} < d < 0$, $f(0) = 0$ so that the sum of all correlations is zero. For the reasons mentioned earlier, this case is of less practical importance.

For $0 < d < \frac{1}{2}$, asymptotic formulas for the covariances and correlations follow from Theorem 2.1 (ii): As $|k| \to \infty$,

$$\gamma(k) \sim c_\gamma(d, \phi, \psi) \, |k|^{2d-1}, \qquad (2.35)$$

where

$$c_\gamma(d, \phi, \psi) = \frac{\sigma_\epsilon^2}{\pi} \frac{|\psi(1)|^2}{|\phi(1)|^2} \Gamma(1 - 2d) \sin \pi d \qquad (2.36)$$

and

$$\rho(k) \sim c_\rho(d, \phi, \psi) \, |k|^{2d-1}, \qquad (2.37)$$

where

$$c_\rho(d, \phi, \psi) = \frac{c_\gamma(d, \phi, \psi)}{\int_{-\pi}^{\pi} f(\lambda) d\lambda}. \qquad (2.38)$$

To obtain explicit formulas for all covariances is slightly more difficult, except in the case of a fractional ARIMA$(0, d, 0)$ process. In this simple case, the form of the covariances follows directly from a formula in Gradshteyn and Ryzhik (1965, p. 372):

$$\gamma(k) = \sigma_\epsilon^2 \frac{(-1)^k \Gamma(1 - 2d)}{\Gamma(k - d + 1)\Gamma(1 - k - d)}. \qquad (2.39)$$

The correlations are equal to

$$\rho(k) = \frac{\Gamma(1-d)\Gamma(k+d)}{\Gamma(d)\Gamma(k+1-d)}. \tag{2.40}$$

Because $\Gamma(k+a)/\Gamma(k+b)$ is approximately equal to k^{a-b} for large k, the asymptotic behavior of $\rho(k)$ is given by

$$\rho(k) \sim \frac{\Gamma(1-d)}{\Gamma(d)}|k|^{2d-1} \ (|k| \to \infty). \tag{2.41}$$

For a general fractional ARIMA(p, d, q) process, the covariances and correlations can be obtained as follows: As we remarked above, X_t is obtained by passing a fractional ARIMA$(0, d, 0)$ process X_t^* through the linear filter

$$\beta(B) = \sum_{j=0}^{\infty} \beta_j B^j = \phi(B)\psi^{-1}(B).$$

Denote by $\gamma^*(k)$ the covariances of X_t^*. In a first step, we calculate the coefficients β_j. This is done by matching the powers of $\phi(B)\psi^{-1}(B)$ with those of $\beta(B)$. For instance, if

$$\phi(B) = 1$$

and

$$\psi(B) = 1 + \psi B,$$

then one obtains

$$\phi(B)\psi^{-1}(B) = 1 - \psi B + \psi^2 B^2 - \ldots$$

Thus,

$$\beta_j = (-1)^j \psi^j.$$

In a second step, the covariances $\gamma(k)$ of the resulting fractional ARIMA(p, d, q) process X_t are obtained from $\beta(B)$ and the covariances $\gamma^*(k)$ by

$$\gamma(k) = \sum_{j,l=0}^{\infty} \beta_j \beta_l \gamma^*(k+j-l). \tag{2.42}$$

We conclude this section by noting some other properties that follow easily from the definition: A fractional ARIMA$(0, d, 0)$ process can be represented as an infinite moving average or an infinite autoregressive process with coefficients that can be given explicitly (Hosking 1981):

Proposition 2.2 Let X_t be a fractional ARIMA(0,d,0)-process with $-\frac{1}{2} < d < \frac{1}{2}$. Then

(i) the following infinite autoregressive representation holds:

$$\sum_{k=0}^{\infty} \pi_k X_{t-k} = \epsilon_t, \qquad (2.43)$$

where ϵ_t $(t = 1, 2, ...)$ are independent identically distributed random variables and

$$\pi_k = \frac{\Gamma(k - d)}{\Gamma(k + 1)\Gamma(-d)}. \qquad (2.44)$$

For $k \to \infty$ we have

$$\pi_k \sim \frac{1}{\Gamma(-d)} k^{-d-1}. \qquad (2.45)$$

(ii) The following infinite moving average represenation holds:

$$X_t = \sum_{k=0}^{\infty} a(k)\epsilon_{t-k} \qquad (2.46)$$

where ϵ_t $(t = 1, 2, ...)$ are independent identically distributed random variables and

$$a(k) = \frac{\Gamma(k + d)}{\Gamma(k + 1)\Gamma(d)}. \qquad (2.47)$$

For $k \to \infty$ we have

$$a(k) \sim \frac{1}{\Gamma(d)} k^{d-1}. \qquad (2.48)$$

These results are especially useful if we want to predict future observations from an (almost) infinite past. For predictions from a finite past, the following result gives an explicit formula for the best linear predictor (Hosking 1981):

Proposition 2.3 *Let X_t be a fractional $ARIMA(0, d, 0)$ process with $-\frac{1}{2} < d < \frac{1}{2}$ and let*

$$\hat{X}_t = \sum_{j=1}^{k} \beta_{kj} X_{t-j}$$

be the best linear predictor of X_t given $X_{t-1}, ..., X_{t-k}$. This means that the coefficients β_{jk} minimize the expected squared prediction error $E[(X_t - \hat{X}_t)^2 | X_{t-1}, ..., X_{t-k}]$. Then

$$\beta_{kj} = -\binom{k}{j} \frac{\Gamma(j - d)\Gamma(k - d - j + 1)}{\Gamma(-d)\Gamma(k - d + 1)}. \qquad (2.49)$$

In particular the partial correlation coefficient β_{kk} is equal to

$$\beta_{kk} = \frac{d}{k - d}. \tag{2.50}$$

For $j, k \to \infty$ and $j/k \to 0$ we have

$$\beta_{kj} \sim \frac{1}{\Gamma(-d)} j^{-d-1}. \tag{2.51}$$

Again it is slightly more complicated to obtain these results for fractional ARIMA(p, d, q) models with non-zero p or q. The principle of how to obtain the coefficients of the best linear unbiased prediction, the infinite moving average, and the autoregressive representations respectively, is, however, the same as for ARMA processes. The partial correlation coefficients are obtained explicitly from the Durbin-Levison algorithm. To obtain α_k ($k = 1, 2, ...$) in the infinite moving average represenation, we expand

$$(1 - B)^{-d} \phi^{-1}(B) \psi(B)$$

in a (formal) power series.

CHAPTER 3

Limit theorems

3.1 Introduction

In this chapter we discuss some limit theorems that are useful for statistical inference. There is an extensive literature on limit theorems for long-memory processes. Here we mainly focus on results that will be needed later.

3.2 Gaussian and non-gaussian time series with long memory

The simplest time series models are Gaussian processes. Their distribution is fully determined by the expected value and the covariances. Methods of inference can therefore be obtained by restricting attention to the first two moments only.

In general, it is unlikely that an observed time series is exactly Gaussian. Good statistical inference procedures should therefore remain valid approximately, even if the actual process deviates from this ideal model. We will see later, at least partially, how such methods may be obtained (Chapter 7). A fully developed theory of robustness for long-memory processes is not known, however, at the present stage. Instead, several authors consider more general conditions under which limit theorems derived for the Gaussian case remain to hold. We consider two types of generalizations of Gaussian time series:

1. *One-dimensional transformations* of a Gaussian time series: Let X_t be a Gaussian process. We define a new process by

$$Y_t = G(X_t), \tag{3.1}$$

where G is a sufficiently regular function.

2. *Linear processes*: Let ϵ_t, $(t = ..., -2, -1, 0, 1, 2, ...)$ be a sequence of independent (not necessarily normal) variables and

let $a(k)$, $(k = ..., -2, -1, 0, 1, 2, ...)$ be a sequence of real numbers. We define a linear process by

$$Y_t = \sum_{k=-\infty}^{\infty} a(k)\epsilon_{t-k}. \tag{3.2}$$

Let us first look at generalization 1: In order that probabilites can be calculated, the function G must be sufficiently regular. In particular, G is assumed to be measurable. Also, we assume that the first two moments, $E[G(X_t)]$ and $E[G^2(X_t)]$, exist and are finite. Without loss of generality we can assume that $E[G(X_t)] = 0$. This can always be achieved by subtracting the expected value $E[G(X_t)]$ from $G(X_t)$.

The limiting behavior of linear sums, quadratic forms, and higher order polynomial forms in $G(X_t)$ turns out to be essentially characterized by the corresponding limiting behavior for Hermite polynomials $H_k(X_t)$. Hermite polynomials are defined by:

Definition 3.1 *The kth Hermite polynomial $H_k(x)$ is equal to*

$$H_k(x) = (-1)^k e^{x^2/2} [\frac{d^k}{dx^k} e^{-x^2/2}]. \tag{3.3}$$

For instance, the first three Hermite polynomials are $H_1(x) = x$, $H_2(x) = x^2 - 1$, and $H_3(x) = x^3 - x$. A useful property of Hermite polynomials is that they build an orthogonal basis in the following sense: Let X be a standard normal random variable. Let $\mathcal{G}$ be the set of functions G with $E[G(X)] = 0, E[G^2(X)] < \infty$ and define the scalar product

$$< G, F > = E[G(X)F(X)]. \tag{3.4}$$

Then

$$< H_k, H_k > = k!, \tag{3.5}$$

for all $k \neq r$

$$< H_k, H_r > = 0, \tag{3.6}$$

and every function $G \in \mathcal{G}$ can be written as

$$G(X) = \sum_{k=0}^{\infty} a_k \frac{1}{k!} H_k(X). \tag{3.7}$$

with *Hermite coefficients*

$$a_k = < G, H_k > . \tag{3.8}$$

Equality is to be understood in the sense that

$$S_K = \sum_{k=0}^{K} a_k \frac{1}{k!} H_k(X) \tag{3.9}$$

converges to $G(X)$ in the L^2 norm $< ., . >$, as K tends to infinity. In particular, (3.4), (3.5), and (3.6) imply

$$\text{var}(G(X)) = \sum_{k=0}^{\infty} \frac{a_k^2}{k!}. \tag{3.10}$$

We will see that in most limit theorems for long-memory processes of the form $G(X_t)$ $(t = 1, 2, ...)$, the only contribution to the limiting distribution comes from the lowest order Hermite polynomial in (3.7) for which the coefficient a_k is not zero. We therefore define the so-called Hermite rank:

Definition 3.2 $G \in \mathcal{G}$ has Hermite rank m, if $a_k = 0$ for $k < m$ and $a_m \neq 0$.

Let us now turn to the second generalization. We assume that the observed process Y_t can be written in the form (3.2). We will assume that Y_t has finite variance. This means that we assume $\sigma_\epsilon^2 = \text{var}(\epsilon_t) < \infty$ and

$$\sum_{k=-\infty}^{\infty} a^2(k) < \infty. \tag{3.11}$$

Equation (3.2) means that Y_t is obtained by passing the sequence of independent observations ϵ_t through the linear filter with coefficients $a(k)$. In applications, it is usually not plausible that Y_t should depend on future values of ϵ_t. The summation in (3.2) is therefore often restricted to $k \geq 0$. That is, one assumes the process to be "causal" in the sense that a one-sided moving representation

$$Y_t = \sum_{k=0}^{\infty} a(k) \epsilon_{t-k} \tag{3.12}$$

exists.

3.3 Limit theorems for simple sums

We consider the limiting distribution of appropriately normalized sums

$$S_n = \sum_{k=1}^{n} Y_t. \tag{3.13}$$

A general result on the limiting behavior of partial sums was given in Theorem 2.3. Essentially it says that the only limiting processes one can obtain by normalizing partial sums S_{nu} appropriately, are self-similar processes. For deriving most of the statistical inference procedures, it will be sufficient to consider the special case $u = 1$. We therefore restrict attention to (3.13).

Suppose that $Y_t = G(X_t)$, where X_t is a stationary process with long-range correlations and $G \in \mathcal{G}$. Without loss of generality, we may assume that X_t has a mean of zero and a standard deviation of 1. The following limit theorem follows from Taqqu (1975, 1979) and Breuer and Major (1983) (see also Dobrushin 1979, Dobrushin and Major 1979, Giraitis and Surgailis 1985):

Theorem 3.1 *Let $G \in \mathcal{G}$ be of Hermite rank m. Then the following holds:*

(i) If $\frac{1}{2} < H < 1 - 1/(2m)$, then

$$\sigma_S^2 = \lim_{n \to \infty} n^{-1} \sum_{j,l=1}^{n} E[G(X_j)G(X_l)] \tag{3.14}$$

exists and

$$S_n^* = n^{-\frac{1}{2}} S_n \to_d S, \tag{3.15}$$

where S is a normal random variable with zero mean and variance σ_S^2.

(ii) If $1 - 1/(2m) < H < 1$, then

$$S_n^* = n^{-1-m(H-1)} S_n \to_d \sqrt{c_m} \frac{a_m}{m!} Z_m, \tag{3.16}$$

where

$$c_m = \frac{2c_\gamma^m m!}{(1 - m(2 - 2H))(2 - m(2 - 2H))} \tag{3.17}$$

and Z_m is a nondegenerate random variable.

The exact distribution of Z_m is rather complicated in general. It is the marginal distribution of a so-called Hermite process $Z_m(u)$ of order m at time $u = 1$. For $m = 1$, Z_m is a standard normal random variable. A formula for the characteristic function of Z_m for $m = 2$ is given in Taqqu (1975). We will restrict attention mostly to the case $m = 1$. For probabilistic characterizations of the general case $m \geq 1$ and more technical details, we refer the interested reader to Taqqu (1975, 1979), Dobrushin (1979), and Dobrushin and Major (1979).

Most important in Theorem 3.1 is the distinction of two types of behavior: For $1 - 1/(2m) < H < 1$, S_n has to be divided by

n^β with $\beta > 1/2$, in order to obtain a nondegenerate limiting distribution. Moreover, the limiting distribution is not necessarily normal. This is in contrast to sums of random variables with summable correlations. If the correlations are summable, then the normalizing factor is $n^{\frac{1}{2}}$ and the limiting distribution is normal.

The situation is very different for $\frac{1}{2} < H < 1 - 1/(2m)$. There, S_n has to be divided by the usual standardization $n^{\frac{1}{2}}$, and we obtain a Gaussian limit. Thus, essentially we are in the same situation as under independence or short-range dependence.

A simple heuristic consideration shows why there is such a distinct contrast between $H < 1 - 1/(2m)$ and $H > 1 - 1/(2m)$. The essential point is that if the covariance between X_j and X_{j+k} is $\gamma(k)$, then the corresponding covariance for the mth Hermite polynomial is given by (see, e.g., Rozanov 1967):

$$\gamma_m(k) = \text{cov}(H_m(X_j), H_m(X_{j+k})) = m! \gamma^m(k). \tag{3.18}$$

In order that S_n has rate of convergence $n^{-\frac{1}{2}}$, the sum of all covariances $\gamma_m(k)$ must be finite. Thus, we need

$$\sum_{k=-\infty}^{\infty} \gamma^m(k) < \infty.$$

This is the case if $H < 1 - 1/(2m)$. On the other hand, for $H > 1 - 1/(2m)$,

$$\sum_{k=-\infty}^{\infty} \gamma^m(k) = \infty,$$

so that the rate of convergence of S_n is slower than $n^{-\frac{1}{2}}$. The case $H = 1 - 1/(2m)$ would have to be treated separately. We omit it here, as it is only of marginal interest.

The second remarkable result in Theorem 3.1 is that for $1 - 1/(2m) < H < 1$, only the first non-zero term in the Hermite expansion

$$G(X) = \sum_{k=m}^{\infty} \frac{a_k}{k!} H_k(X)$$

determines the asymptotic distribution. The asymptotic distribution of $n^{-1-m(H-1)} S_n$ is exactly the same as the asymptotic distribution of $n^{-1-m(H-1)} S_{m,n}$ where

$$S_{m,n} = \frac{a_m}{m!} \sum_{j=1}^{n} H_m(X_j). \tag{3.19}$$

The reason is that for $m' > m$,

$$\lim_{n \to \infty} \sum_{k=-(n-1)}^{n-1} \gamma^{m'}(k) \Big/ \sum_{k=-(n-1)}^{n-1} \gamma^{m}(k) = 0.$$

Therefore, the variance of $S_{m',n}$ diverges slower to infinity than $n^{1+m(H-1)}$ and $n^{-1-m(H-1)}S_{m',n}$ tends to zero in probability.

A useful generalization of Theorem 3.1 is obtained by considering transformations of X_t that depend on an index v, where v is in a compact interval. We define

$$Y_t = G(X_t, v). \qquad (3.20)$$

Then

$$S_n(v) = \sum_{j=1}^{n} G(X_i, v), \quad n = 1, 2, \ldots \qquad (3.21)$$

is a sequence of stochastic processes. Under certain additional regularity conditions on G, one can show that the stochastic process $S_n(v)$, normalized appropriately, converges weakly to a well-defined limiting process. We will not need this result in its full generality. Instead, we consider two special cases that are of interest in statistical applications:

1. "Well behaved" continuous functions G with $v \in [-V, V]$;

2. Indicator funtions $G(X_t, v) = 1\{\tilde{G}(X_t) \le v\}$, with $v \in [-\infty, \infty]$.

The following is a consequence of the above theorem and Billingsley (1968, p. 97) (see Beran and Ghosh 1991):

Theorem 3.2 *Let* $0 < V < \infty, x \in R$ *and* $v \in [-V, V]$. *Let* $G(x, v)$ *be continuous and such that for each* $v \in [-V, V]$, $g(x) = G(x, v)$ *is in* $\mathcal{G}$ *and of Hermite rank* $m \ge 1$ *with* mth *Hermite coefficient* $a_m(v)$. *Suppose that there exists a constant* $\gamma > 1$ *and a nondecreasing continuous function* $\psi : [-V, V] \to R$ *such that, for a standard normal random variable* X,

$$E(|G(X, v) - G(X, w)|^2) \le |\psi(v) - \psi(w)|^{\gamma} \qquad (3.22)$$

for all $v, w \in [-V, V]$. *Then the following holds:*

(i) If $\frac{1}{2} < H < 1 - 1/(2m)$, *then*

$$c(v, w) = \lim_{n \to \infty} n^{-1} \sum_{j,l=1}^{n} E[G(X_j, v) G(X_l, w)] \qquad (3.23)$$

exists and

$$S_n^*(v) = n^{-\frac{1}{2}} S_n(v) \qquad (3.24)$$

converges to a zero mean Gaussian process with covariance function $c(v, w)$. The convergence is weak convergence in the space $C[-V, V]$ of continuous functions on $[-V, V]$, equipped with the supremum norm.

(ii) If $1 - 1/(2m) < H < 1$, then

$$S_n^*(v) = n^{-1-m(1-H)} S_n(v) \qquad (3.25)$$

converges (in the same sense as above) to

$$\sqrt{c_m} \frac{a_m(v)}{m!} Z_m \qquad (3.26)$$

where c_m and Z_m are as in Theorem 3.1.

The following theorem is a special case of a result by Dehling and Taqqu (1989):

Theorem 3.3 *Let $\tilde{G}$ be a function in $\mathcal{G}$ and define $G(x, v) = 1\{\tilde{G}(x) \leq v\}$ Assume that, for each $v \in R$, $g(x) = G(x, v)$ is of Hermite rank m. Denote by $D[-\infty, \infty]$ the space of functions defined for all real numbers (including $-\infty, \infty$) that are right-continuous and for which limits from the left exist, equipped with the supremum norm. Let $1 - 1/(2m) < H < 1$. Then*

$$S_n^*(v) = n^{-1-m(H-1)} S_n(v) \qquad (3.27)$$

converges weakly to

$$\sqrt{c_m} \frac{a_m(v)}{m!} Z_m. \qquad (3.28)$$

Note that the limitng processes in Theorems 3.2(ii) and 3.3 are a constant function $\sqrt{c_m(v)}$ times a random variable. Thus, once we observe one observation, we know the whole sample path.

The results dicussed in this Section can be derived for more general processes (Surgailis 1981a, Giraitis and Surgailis 1985, 1986, Avram and Taqqu 1987). In particular, Surgailis (1981a) derived analogous limit theorems for linear processes (3.12).

3.4 Limit theorems for quadratic forms

Quadratic forms play a central role in approximations of the Gaussian maximum likelihood. The likelihood function of a normally distributed random vector $X = (X_1, ..., X_n)^t$ with mean zero and covariance matrix $\Sigma = [\gamma(i - j)]_{i,j=1,...,n}$, is given by

$$g(X_1, ..., X_n) = (2\pi)^{-\frac{n}{2}} |\Sigma|^{-\frac{n}{2}} e^{-\frac{1}{2}X^t \Sigma^{-1} X}. \qquad (3.29)$$

Here, $|\Sigma|$ denotes the determinant of Σ. In statistical applications, Σ is assumed to depend on an unknown parameter vector θ. The likelihood function g is then a function of X and θ. For instance, for fractional Gaussian noise with zero mean and variance σ^2, one can define the parameter vector $\theta = (\sigma^2, H)$ and the covariance matrix is of the form

$$\Sigma_{i,j}(\theta) = \frac{1}{2}\sigma^2(|i - j + 1|^{2H} - 2|i - j|^{2H} + |i - j - 1|^{2H}). \qquad (3.30)$$

The maximum likelihood estimator (MLE) of θ is obtained by maximizing $g(X_1, ..., X_n; \theta)$ with respect to θ. To evaluate g, one needs to invert the $n \times n$-matrix Σ. In an iterative maximization procedure, many evaluations of g have to be performed. For long data sets, inversion of Σ can be time and space (computer memory) consuming. It is therefore often useful to replace Σ^{-1} by a matrix that can be calculated directly without inversion. A simple approximation of Σ^{-1} is obtained by the matrix A with elements

$$\alpha_{j,l} = \alpha_{j-l} = (2\pi)^{-2} \int_{-\pi}^{\pi} \frac{1}{f(\lambda; \theta)} e^{i(j-l)\lambda} d\lambda \qquad (3.31)$$

where f is the spectral density (Whittle 1951, see also Bleher 1981). Based on this approximation, and a suitable parametrization $\theta = (\sigma_\epsilon^2/(2\pi), \eta)$ (see Chapter 5), an approximate MLE of θ is obtained by minimizing the quadratic form

$$Q^*(\eta) = X^t A(\theta^*) X \qquad (3.32)$$

with respect to η, where $\theta^* = (1, \eta)^t$, and setting

$$\hat{\sigma}_\epsilon^2 = 2\pi Q^*(\hat{\eta}). \qquad (3.33)$$

This will be discussed in detail in Chapter 5. Under mild regularity conditions, minimizing $Q^*(\eta)$ is equivalent to finding the solution of

$$\dot{Q}^* = \frac{\partial}{\partial \eta_j} Q^*(\eta) = 0 \ (j = 1, 2, ...). \qquad (3.34)$$

These considerations motivate the following probabilistic problem: Given a long-memory process and a symmetric Toeplitz matrix B, how do we need to standardize the quadratic form

$$Q = X^t B X \qquad (3.35)$$

in order to obtain a nondegenerate limiting distribution, and what is the resulting limit distribution? Recall that B is called a symmetric Toeplitz matrix if B is symmetric with elements $B_{ij} = b_{|i-j|}$.

At first, suppose that X_i is a Gaussian process. The asymptotic distribution of Q follows from Fox and Taqqu (1987) and Avram (1988a).

Theorem 3.4 *Let X_i be a stationary Gaussian process with mean zero and spectral density f. Also, let B be the Toeplitz matrix whose elements b_{i-j} are defined by*

$$b_k = \int_{-\pi}^{\pi} e^{ikx} g(x)dx \qquad (3.36)$$

for some g. Assume

$$\frac{1}{n} Trace(\Sigma B)^2 \to (2\pi)^3 \int_{-\pi}^{\pi} [f(x)g(x)]^2 dx < \infty \qquad (3.37)$$

and that there exists an $\alpha < 1$ and $\beta < 1$ such that

$$\alpha + \beta < \frac{1}{2}, \qquad (3.38)$$

$$f(x) \sim c_f |x|^{-\beta} \text{ as } |x| \to 0 \qquad (3.39)$$

and

$$g(x) \sim c_g |x|^{-\alpha} \text{ as } |x| \to 0, \qquad (3.40)$$

where c_f and c_g are finite constants. Then, under mild additional regularity conditions on f and g,

$$n^{-\frac{1}{2}}[Q - E(Q)] \to_d \sigma_Q^2 Z, \qquad (3.41)$$

where Z is a standard normal random variable and

$$\sigma_Q^2 = 16\pi^3 \int_{-\pi}^{\pi} [f(x)g(x)]^2 dx. \qquad (3.42)$$

Note that in the context of long-memory processes, we consider spectral densities with β in the range $(0, 1)$. In our notation, $\beta = 2H - 1$. Condition (3.38) determines the strength of the long memory. Similarly, (3.39) determines how fast the elements b_{i-j} of the matrix B converge to zero as $|i - j|$ tends to infinity. If $0 < \alpha < 1$, then $b_k \sim c_b k^{\alpha-1}$ as $|k| \to \infty$, where $0 < c_b < \infty$. For $\alpha < 0$, $\sum_{k=-\infty}^{\infty} b_k = 0$ and $\sum_{k=-\infty}^{\infty} |b_k| < \infty$. Condition (3.37) ensures that b_k converges fast enough to zero in order to compensate for the long memory of X_i. If the long memory is strong (β large), then b_k has to converge fast enough to zero in order that the variance of Q remains to be of the order $O(n)$. The mathematical reason is that in order for the variance of Q to be of this order, (3.36) must hold. Finally, it should be noted that the same central limit theorem holds when the sample mean is subtracted from X_i (Fox and Taqqu 1987, Theorem 4). Also note that a different behavior of Q is obtained if $\alpha + \beta > \frac{1}{2}$. Terrin and Taqqu (1990a) showed that in this case, Q has to be divided by $n^{\alpha+\beta}$ and the limiting distribution is non-normal.

A generalization to linear processes (3.6) is given in Giraitis and Surgailis (1990a):

Theorem 3.5 *Let Y_t be defined by (3.6). Denote by χ_4 the fourth cumulant of ϵ_o and let f be the spectral density of Y_t. Then under the same conditions on f as in Theorem 3.4,*

$$n^{-\frac{1}{2}}[Q - E(Q)] \to_d \sigma_Q^2 Z, \tag{3.43}$$

where Z is a standard normal random variable and

$$\sigma_Q^2 = 16\pi^3 \int_{-\pi}^{\pi} [f(x)g(x)]^2 dx + \chi_4[2\pi \int_{-\pi}^{\pi} f(x)g(x)dx]^2. \tag{3.44}$$

In particular, if Y_t is Gaussian, then Theorem 3.4 follows as a special case. Again, for $\alpha + \beta > \frac{1}{2}$, Q has to be divided by $n^{\alpha+\beta}$ and the limiting distribution is non-normal (Terrin and Taqqu 1990a)

Theorem 3.5 can be generalized to a multivariate limit theorem (Beran and Terrin 1994):

Theorem 3.6 *For $l = 0, 1, 2, ...$, let*

$$Q_{n,l} = \sum_{t,s=1}^{n} b_{t-s} Y_{t+ln} Y_{s+ln} = \sum_{t,s=1+ln}^{(1+l)n} b_{t-s} Y_t Y_s. \tag{3.45}$$

Suppose that $l_1, ..., l_p$, $p = 1, 2, ...$, are distinct positive integers. Then, under the assumptions of Theorem 3.5,

$$n^{-\frac{1}{2}}(Q_{n,l_1} - EQ_{n,l_1}, ..., Q_{n,l_p} - EQ_{n,l_p}) \to_d \sigma_Q^2(Z_1, Z_2, ..., Z_p), \tag{3.46}$$

where $Z_1, ..., Z_p$ are independent standard normal random variables and σ_Q^2 is defined in (3.44).

The essential message of this theorem is that if we consider disjoint subseries, then the corresponding quadratic forms are independent.

Other generalizations include limit theorems for forms defined by $Q = \sum_{j,l=1}^{n} b_{j-k} G(X_j, X_l)$, where G is more general than the product $X_j X_l$ (see Giraitis and Surgailis 1986, Avram 1988c, Terrin and Taqqu 1991b), and weak convergence results for partial sums, with summation over $i, j = 1, ..., [nt]$ where t is in a compact interval.

3.5 Limit theorems for Fourier transforms

In the previous section we mentioned an approximate MLE method for estimating an unknown parameter vector θ, based on quadratic forms. Another approach is to estimate θ from the periodogram $I(\lambda)$. In this section, we discuss the asymptotic behavior of the periodogram for long-memory time series. We assume $E(X_i)$ to be known and equal to zero. The results also hold when X_i is replaced by $X_i - \bar{X}$.

For stationary time series with a bounded spectral density, it is well known that, for $\lambda \neq 0$,

$$I(\lambda) \to_d f(\lambda)\xi, \tag{3.47}$$

where ξ is an exponential random variable with mean 1. Moreover, consider a finite number of frequencies $\lambda_{n,1}, ..., \lambda_{n,k}$ with

$$\lambda_{n,j} \to \lambda_j \tag{3.48}$$

and

$$\lambda_j \pm \lambda_{j'} \neq 2\pi l \ (l \in Z) \tag{3.49}$$

for $j \neq j'$. Then

$$[I(\lambda_{n,1}), ..., I(\lambda_{n,k})] \to_d [f(\lambda_1)\xi_1, ..., f(\lambda_k)\xi_k], \tag{3.50}$$

where $\xi_1, ..., \xi_k$ are independent standard exponential random variables. These results can be used to estimate θ via maximum likelihood for independent exponential random variables: $\hat{\theta}$ is the maximum likelihood estimator based on the exponentially distributed independent random variables

$$I(\lambda_{n,j}), \ j = 1, ..., [(n - 1)/2]$$

with means

$$f(\lambda_{n,j}; \theta), \ j = 1, ..., [(n - 1)/2],$$

where

$$\lambda_{n,j} = \frac{2j\pi}{n} \ (j = 1, ..., n^*)$$

are the Fourier frequencies and n^* is the integer part of $(n - 1)/2$. It should be noted, however, that for general (not necessarily Gaussian) linear processes, (3.50) holds only for a fixed number of frequencies, whereas $\hat{\theta}$ is a function of the periodogram at an increasing number of frequencies. Nevertheless, $\hat{\theta}$ turns out to have the same asymptotic distribution as if (3.50) were true for all Fourier frequencies (cf. Chapter 5).

Yajima (1989a) showed that, under certain mild regularity conditions, (3.50) remains to hold for long-memory processes:

Theorem 3.7 *Let X_t be a stationary process with spectral density f. Suppose that there is a positive continuous function f^* : $[-\pi, \pi] \to R_+$ such that the spectral density of ϵ_t can be written as*

$$f(x) = f^*(x)|1 - e^{ix}|^{1-2H} \tag{3.51}$$

with $\frac{1}{2} < H < 1$. Also, assume that X_t has a one-sided infinite moving average representation (3.12) and that (3.48) and (3.49) hold. Then, under some conditions on the cumulants of ϵ_t, (3.50) holds.

Note in particular that X_t does not need to be Gaussian. More generally, Yajima (1989a) showed that, under the assumptions of Theorem 3.7, the limiting behavior of the tapered periodogram

$$I_h(\lambda) = \frac{1}{2\pi n}|d_n(\lambda)|^2 \tag{3.52}$$

is the same as for short-memory processes. Here,

$$d_n(\lambda) = \sum_{t=1}^{n} h(tn^{-1})X_t e^{-i\lambda t}, \tag{3.53}$$

and h is a sufficiently regular real-valued function. Tapering is a common technique to reduce the bias of spectral estimates due to leakage (see, e.g., Priestley 1981, p. 563). Note that, if $h \equiv 1$, then I_h is the usual periodogram.

One of the crucial assumptions in Theorem 3.7 is that none of the frequencies converges to zero. For frequencies that tend to zero, the asymptotic behavior of the periodogram is quite different. Yajima also considered the case $\lambda_{n,j} = 0$ and obtained a different rate of convergence. For Fourier frequencies tending to zero, the following result holds (Künsch 1986a, Hurvich and Beltrao 1992, Robinson 1992a):

Theorem 3.8 *Let X_t be a stationary process with long memory and a spectral density f as in Theorem 3.7. For a fixed integer j, let $\lambda_j = 2\pi j/n$ be the jth Fourier frequency. Also, define the normalized periodogram*

$$I^*(\lambda_j) = \frac{I(\lambda_j)}{f(\lambda_j)}. \tag{3.54}$$

Then the following results hold.

(i) The asymptotic bias of $I^*(\lambda_j)$ is equal to

$$\lim_{n\to\infty} E[I^*(\lambda_j)] = 2\Delta_j(H, -1), \qquad (3.55)$$

where

$$\Delta_j(H, u) = \frac{1}{\pi} \int_{-\infty}^{\infty} \frac{\sin^2 \frac{x}{2}}{(2\pi j - x)(2\pi j + ux)} |\frac{x}{2\pi j}|^{1-2H} dx. \qquad (3.56)$$

(ii) The asymptotic variance of $I(\lambda_j)$ is of the order

$$\mathrm{var}[I(\lambda_j)] = O(n^{4H-2}). \qquad (3.57)$$

(iii) For $j \neq j'$, the normalized periodogram ordinates $I^*(\lambda_j), I^*(\lambda_{j'})$ are asymptotically correlated.

(iv) If X_t is Gaussian, then

$$I^*(\lambda_j) \to_d \eta_j, \qquad (3.58)$$

where η_j is defined by

$$\eta_j = [\frac{1}{2}\Delta_j(H, -1) - \Delta_j(H, 1)]Z_1^2 + [\frac{1}{2}\Delta_j(H, -1) + \Delta_j(H, 1)]Z_2^2 \qquad (3.59)$$

with independent standard normal random variables Z_1, Z_2.

Numerical calculations in Hurvich and Beltrao (1992) show that, for $H > \frac{1}{2}$, the asymptotic bias $\Delta_j(H, 1)$ is large for small values of j. For instance, for the first Fourier frequency and $H = 0.99$, one obtains $\Delta_1(0.99, 1) = 1.14$. On the other hand, the bias seems to approach 1 rather quickly as j increases. Also, the bias decreases as H approaches $\frac{1}{2}$. The second result in Theorem 3.8 means that, for frequencies tending to zero, the periodogram fluctuates much more than for short-memory processes, and needs to be normalized accordingly. The third result is also in contrast to results for short-memory processes. There, periodogram ordinates at different Fourier frequencies are asymptotically uncorrelated. Finally, for a Gaussian process, Theorem 3.8(iv) implies that, for $H > \frac{1}{2}$, the asymptotic distribution of the normalized peridogram is no longer exponential. Note that, for $H = \frac{1}{2}$, $\Delta_j(\frac{1}{2}, 1) = 1$ and $\Delta_j(\frac{1}{2}, -1) = 0$, so that the well-known result for short-memory processes follows.

In contrast to the periodogram, the integrated periodogram

$$\hat{F}_n(\lambda) = \frac{2\pi}{n} \sum_{j=1}^{n_\lambda} I(\lambda_j) \qquad (3.60)$$

with $n_\lambda = [n\lambda/(2\pi)]$ converges to its population counterpart $F(\lambda)$ even if n_λ tends to zero. Robinson (1991f) proved the following result.

Theorem 3.9 *Let X_t be a linear process* (3.12) *with suitable regularity conditions on ϵ_t and such that* (2.2) *holds. If*

$$\frac{1}{m} \to 0 \qquad (3.61)$$

and

$$\frac{m}{n} \to 0 \qquad (3.62)$$

then

$$\frac{\hat{F}_n(\lambda_m)}{F_n(\lambda_m)} \to 1 \qquad (3.63)$$

in probability.

Assuming some additional regularity conditions, Robinson also gives a rate at which the asymptotic limit is achieved.

## CHAPTER 4

Estimation of long memory: heuristic approaches

4.1 Introduction

The phenomenon of long memory was observed in applications long before appropriate stochastic models were known. Several heuristic methods to estimate the long-memory parameter H were suggested. Best known is the R/S statistic, which was first proposed by Hurst (1951) in a hydrological context. Other methods include the log-log correlogram, the log-log plot of $\mathrm{var}(\bar{X}_n)$ versus n, the semivariogram, and least squares regression in the spectral domain. These methods are mainly useful as simple diagnostic tools. They are less suitable for statistical inference, as more efficient and more flexible methods exist. Also, for most of these methods, it is not easy to obtain simple confidence intervals. This makes it difficult to interpret results in an objective way. Nevertheless, as exploratory tools, these methods deserve to be discussed in some detail.

4.2 The R/S statistic

Let $Q = Q(t, k) = R(t, k)/S(t, k)$ be the R/S statistic defined in Chapter 1. To estimate the long-memory parameter, the logarithm of Q is plotted against k. For each k, there are $n - k$ replicates.

For illustration, consider the following examples: (1) The Nile River minima ($n = 660$, see Chapter 1) and (2) 660 simulated independent standard normal observations. The logarithm of Q versus $\log k$ is displayed in Figures 4.1 and 4.2, respectively, for $k = 10l$ ($l = 1, ..., 20$) and $t = 60m + 1$ ($m = 1, 2, ...$). We observe that with increasing k, the values of the R/S statistic stabilize around a straight line with a slope approximately equal to 0.936. On the other hand, for the iid data, the R/S statistics is scattered around a straight line with a slope of about 0.543. These observations can

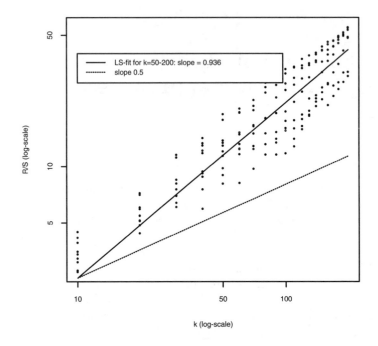

Figure 4.1. *R/S plot for Nile River minima.*

be explained by the following theorems (Mandelbrot 1975):

Theorem 4.1 *Let X_t be such that X_t^2 is ergodic and $t^{-\frac{1}{2}} \sum_{s=1}^{t} X_s$ converges weakly to Brownian motion as t tends to infinity. Then, as $k \to \infty$,*

$$k^{-\frac{1}{2}} Q \to_d \xi, \tag{4.1}$$

where ξ is a nondegenerate random variable.

The assumptions of Theorem 4.1 hold for most common short-memory processes. In a simplified way one may say that whenever the central limit theorem holds, the statistic $k^{-\frac{1}{2}} Q$ converges to a well-defined random variable. For statistical applications it means that in the plot of $\log Q$ against $\log k$, the points should ultimately (for large values of k) be scattered randomly around a straight line with slope $\frac{1}{2}$. In contrast, the following behavior is obtained for

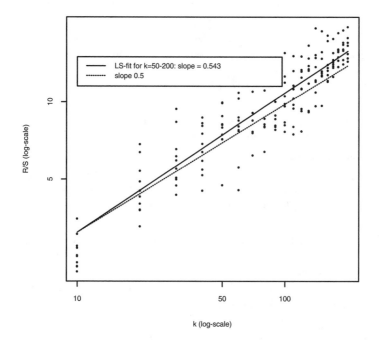

Figure 4.2. R/S-plot for iid normal observations.

long-memory processes:

Theorem 4.2 *Let X_t be such that X_t^2 is ergodic and $t^{-H} \sum_{s=1}^{t} X_s$ converges weakly to fractional Brownian motion as t tends to infinity. Then, as $k \to \infty$,*

$$k^{-H} Q \to_d \xi, \qquad (4.2)$$

where ξ is a nondegenerate random variable.

Thus, for long-memory processes, the points in the R/S plot should be scattered randomly around a straight line with slope $H > \frac{1}{2}$, for sufficiently large lags k. Based on this result, Figure 4.1 suggests that the Nile River data have strong long memory.

A nice property of the R/S statistic is that its asymptotic behavior remains unaffected by long-tailed marginal distributions, in the following sense (Mandelbrot 1975):

Theorem 4.3 *Let X_t be iid random variables with $E(X_t^2) = \infty$ and such that they are in the domain of attraction of stable distributions with index $0 < \alpha < 2$. Then the conclusion of Theorem 4.1 holds.*

Thus, even if X_t has a long-tailed marginal distribution, the R/S statistic still reflects the independence in that the asymptotic slope in the R/S plot remains to be $\frac{1}{2}$.

Based on theorems 4.1 to 4.3, the R/S method can be summarized as follows:

1. Calculate Q for all possible (or for a sufficient number of different) values of t and k.

2. Plot $\log Q$ against $\log k$.

3. Draw a straight line $y = a + b \log k$ that corresponds to the "ultimate" behavior of the data. The coefficients a and b can be estimated, for instance, by least squares or "by eye". Set $\hat{H}$ equal to $\hat{b}$.

The following difficulties arise: How do we decide from which k on the "ultimate behavior" starts? How uncertain is the estimate of H? In particular, for finite samples, the distribution of Q is neither normal nor symmetric. This makes estimation by eye more difficult. Also, it raises the question of whether least squares regression is appropriate. The exact distribution of Q seems to be difficult to derive and depends on the actual distribution of the data generating process. The values of Q for different time points t and lags k are not independent from each other. The exact description of the dependence would be very complicated and possibly model-dependent. Finally, for large lags k, only very few values of Q can be calculated. Because of these problems, it seems difficult to define a fully "automatic" R/S methodology, and to derive results on statistical inference based on the method.

Example: Consider the R/S plot of the Nile River data (Figure 4.1). The difficulty of choosing the cut-off point is illustrated by the estimates of H obtained by fitting a least squares line for $k \geq 10$, $k \geq 40$, $k \geq 50$, and $k = 100$. The estimates are 0.856, 0.910, 0.972, and 1.174 respectively. Note in particular that the last value is outside the range $0 < H < 1$.

The counterpart to Theorem 4.3 is the lack of robustness with respect to the following departures from stationarity (Bhattacharya, Gupta, and Waymire 1983):

Theorem 4.4 *Let ϵ_t be an ergodic stochastic process with zero mean, variance 1, summable correlations, and such that it is in the* *domain of attraction of Brownian motion. Let $\mu(t)$ be a deterministic function. Define*

$$\mu_k(t) = \sum_{j=1}^{t} \mu(j) - \frac{t}{k} \sum_{j=1}^{k} \mu(j) \qquad (4.3)$$

and

$$\Delta_k = \frac{1}{\sqrt{k}}[\max_t \mu_k(t) - \min_t \mu_k(t)]. \qquad (4.4)$$

Suppose the observed process is given by

$$X_t = \mu(t) + \epsilon_t. \qquad (4.5)$$

Then the following two statements are equivalent

(i) There exists an $H_\mu > \frac{1}{2}$ such that, as $k \to \infty$,

$$k^{-H_\mu}Q \to_d \xi, \qquad (4.6)$$

where ξ is almost surely not zero.

(ii)

$$k^{\frac{1}{2}-H_\mu}\Delta_k \to c > 0, \qquad (4.7)$$

as $k \to \infty$ where $H_\mu > \frac{1}{2}$.

Also, if (i) or (ii) holds, then

$$k^{-H_\mu}Q \to c \qquad (4.8)$$

in probability.

Theorem 4.5 *Under the same assumptions as above, but*

$$\Delta_k = o(1), \qquad (4.9)$$

the asymptotic limit (in distribution) of the R/S statistic is given by

$$k^{-\frac{1}{2}}Q \to_d \xi, \qquad (4.10)$$

where $0 < \xi < \infty$ with probability 1.

Note that equation (4.17) implies that Δ_k tends to infinity at the rate $k^{H_\mu-\frac{1}{2}}$. This happens for slowly decaying trends. In constrast, equation (4.18) means that Δ_k converges to zero. This is the case if the trend decays to zero fast. The results imply that the R/S estimate can be misleading if there is a slowly decaying trend. The values of Q are the same as if there were long memory in the data, whereas in reality the stationary part of X_t has short memory only and the trend disappears asymptotically. This is in

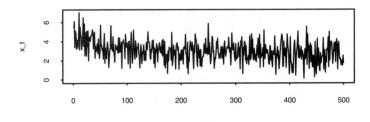

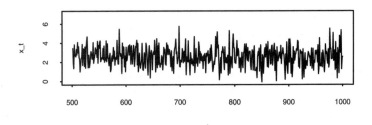

Figure 4.3. *Independent observations plus a slowly decaying trend.*

contrast to certain maximum likelihood based methods which we will discuss later.

Example: Consider $X_t = 10t^{-\beta} + \epsilon_t$ with $0 < \beta < 1$ and ϵ_t iid standard normal. Two cases can be distinguished: $\beta < \frac{1}{2}$ and $\beta > \frac{1}{2}$. In the first case, (4.17) holds with $H_\mu = 1 - \beta$. Thus, the R/S-based estimate of H will tend to the value $H_\mu = 1 - \beta > \frac{1}{2}$. This "long memory" is caused by the slowly decaying trend $\mu(t)$ only. An illustration with $\beta = 0.1$ is given in Figures 4.3 and 4.4. In the second case, the trend decays sufficiently fast in order for (4.19) to hold. Therefore, the R/S-based estimate of the long-memory parameter of X_i is asymptotically equal to $\frac{1}{2}$. An illustration with $\beta = 0.9$ is given in Figures 4.5 and 4.6.

In spite of its deficiencies, the R/S plot is useful. It gives a first aproximate idea about the long-term behavior of the data. For

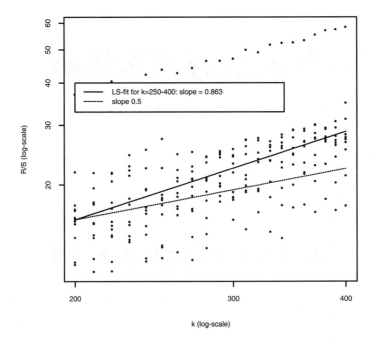

Figure 4.4. *R/S plot for independent observations plus a slowly decaying trend shown in Figure 4.3.*

more detailed results on the $R/S-$statistic we refer the interested reader to for example Feller (1951), Mandelbrot (1972, 1975), and Mandelbrot and Wallis (1969c).

4.3 The correlogram and partial correlations

The plot of the correlations (*correlogram*)

$$\hat{\rho}(k) = \frac{\hat{\gamma}(k)}{\hat{\gamma}(0)} \tag{4.11}$$

against the lag k, as well as the plot of *partial correlations* against k, are standard methods in time series analysis. As a simple rule of thumb one draws two horizontal lines at the levels $\pm 2/\sqrt{n}$. Correlations outside of this band are considered significant at the level

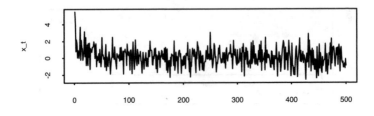

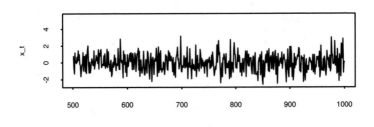

Figure 4.5. *Independent observations plus a quickly decaying trend.*

of significance 0.05. The justification is given by the limiting distribution of the sample correlation at lag k. If the true correlations are zero (see, e.g., Priestley 1981, p. 340), then under mild regularity conditions $\sqrt{n}\hat{\rho}(k)$ are asymptotically independent standard normal random variables. Note, however, that if X_t is not uncorrelated, then the asymptotic variance of $\hat{\rho}(k)$ depends on $\rho(k)$ and possibly on higher moments of X_t (see Priestley 1981, p. 332), and the sample correlations at different lags are not uncorrelated anymore. This makes it more difficult to interpret the correlogram correctly in practice. For example, often several neighbouring sample correlations may be large, although only one of them is nonzero in reality. Nevertheless, the correlogram is a useful diagnostic plot for short-memory processes. For example, it can be used for a preliminary identification of the order q of a moving average pro-

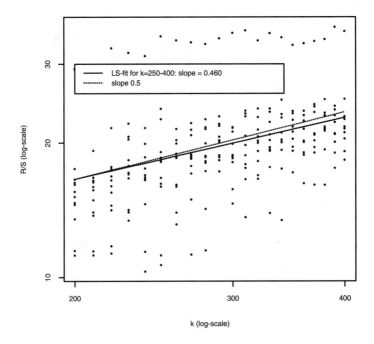

Figure 4.6. *R/S plot for independent observations plus a quickly decaying trend shown in Figure 4.5.*

cess $X_t = \sum_{j=1}^{q} \psi_j \epsilon_{t-j} + \epsilon_t$, as the only non-zero correlations are $\rho_o, \rho_1, ..., \rho_q$. Analogously, partial correlations may be used to identify the order p of an autoregressive process. This is discussed in detail, for example, in Box and Jenkins (1970).

How useful are the sample correlations and partial correlations for detecting long memory? Long memory is characterized by a slow decay of the correlations proportional to k^{2H-2} for some $\frac{1}{2} < H < 1$. A plot of the sample correlations should therefore exhibit this slow decay. Figures 4.7a and 4.7c show the correlogram for simulated fractional ARIMA$(0,d,0)$ series with $H = d + \frac{1}{2} = 0.6$ and $H = 0.9$. For comparison, the correlograms for simulated series of the AR(1) processes with the same lag-1 correlation are given in Figures 4.8a and 4.8c. For $H = 0.9$, the slow decay of the

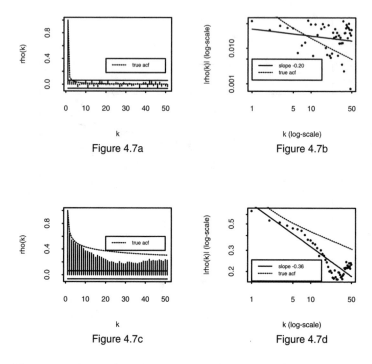

Figure 4.7a

Figure 4.7b

Figure 4.7c

Figure 4.7d

Figure 4.7. *Sample autocorrelations of a simulated fractional ARIMA(0,0.1,0) series (Figures 4.7a and b) and fractional ARIMA(0,0.4,0) series (Figures 4.7c and d) of length 1000.*

sample correlations is clearly visible. The slow decay is much less obvious for $H = 0.6$. It is difficult to tell whether ultimately the correlations follow a hyperbolic curve proportional to k^{2H-2} (for some $\frac{1}{2} < H < 1$), or an exponential curve c^k (for some $0 < c < 1$). Even more difficult would be to distinguish different values of H. A second difficulty is that the definition of long memory only implies that the decay of the correlations is slow. The absolute values of the correlations can be small. Therefore, $\pm 2/\sqrt{n}$ limits will often not recognize that there is dependence in the data. A third difficulty is that long memory is an asymptotic notion. We therefore have to look at the correlations at high lags, However, correlations at high lags can not be estimated reliably.

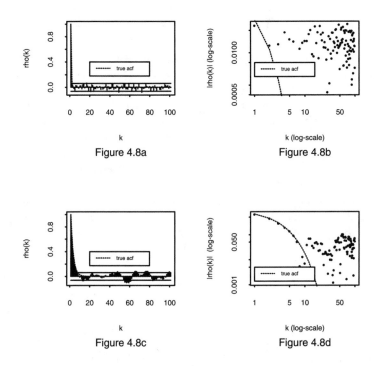

Figure 4.8a

Figure 4.8b

Figure 4.8c

Figure 4.8d

Figure 4.8. *Sample autocorrelations of simulated AR(1) series of length 1000 with the same lag-1 autocorrelation as the processes used for Figures 4.7a,b (Figures 4.8a,b) and Figures 4.7c,d (Figures 4.8c,d) respectively.*

A more suitable plot can be obtained by taking the logarithm on both sides, i.e., by plotting

$$\log|\rho(k)| \text{ against } \log k. \qquad (4.12)$$

If the asymptotic decay of the correlations is hyperbolic, then for large lags, the points in the plot should be scattered around a straight line with negative slope approximatley equal to $2H - 2$. In contrast, for short-memory processes, the log-log correlogram should show divergence to minus infinity at a rate that is at least exponential. For illustration, Figures 4.7b and 4.7d as well as 4.8b and d show the log-log correlogram for simulated fractional ARIMA(0,d,0) series with $H = 0.6$ and $H = 0.9$ and for AR(1) series with the same lag-1 correlation respectively. The plots illustrate that the correlogram in log-log coordinates is mainly useful in cases where long-range dependence is strong or for very long time series. For relatively short series or if H is close to $\frac{1}{2}$, it is very difficult to estimate the ultimate decay of $\rho(k)$ and to decide whether there is long memory in the data, by looking at the log-log correlogram only. Essentially, the same remarks as for the R/S plot apply here regarding the difficulties of interpreting the plot.

Analogous comments also apply to the partial correlations. For long-memory processes, partial correlations decay at the hyperbolic rate $k^{-H-\frac{1}{2}}$. For short-memory processes, the partial correlations are bounded by an exponentially decaying bound. The log-log plot of partial correlations can be interpreted in an analogous way as the log-log correlogram. The difficulties remain the same.

In conclusion, we can say that methods based on ordinary and partial correlation plots are useful for a first heuristic analysis of the data. Some caution is necessary, however, when interpreting such plots, because of the messy sampling properties. A considerable amount of skill and experience is needed to avoid potential pitfalls.

4.4 Variance plot

As we have seen in Chapter 1, one of the striking properties of long-memory processes is that the variance of the sample mean converges slower to zero than n^{-1}. From Theorem 2.2 we have

$$\text{var}(\bar{X}_n) \approx cn^{2H-2}, \tag{4.13}$$

where $c > 0$. This suggests the following method for estimating H:

1. Let k be an integer. For different integers k in the range $2 \leq k \leq n/2$, and a sufficient number (say m_k) of subseries of length k, calculate the sample means $\bar{X}_1(k), \bar{X}_2(k), ..., \bar{X}_{m_k}(k)$ and the overall mean

$$\bar{X}(k) = m_k^{-1} \sum_{j=1}^{m_k} \bar{X}_j(k). \tag{4.14}$$

2. For each k, calculate the sample variance of the sample means $\bar{X}_j(k)$ $(j = 1, ..., m_k)$:

$$s^2(k) = (m_k - 1)^{-1} \sum_{k=1}^{m_k} (\bar{X}_j(k) - \bar{X}(k))^2. \tag{4.15}$$

3. Plot $\log s^2(k)$ against $\log k$.

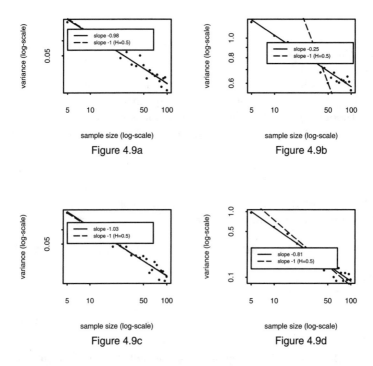

Figure 4.9. *Variance of sample mean vs. sample size (log-log scale) for the series used in Figures 4.7a,b (Figure 4.9a), Figures 4.7c,d (Figure 4.9b), Figures 4.8a,b (Figure 4.9c), and Figures 4.8c,d (Figure 4.9d).*

For large values of k, the points in this plot are expected to be scattered around a straight line with negative slope $2H - 2$. In the case of short-range dependence or independence, the ultimate slope is $2\frac{1}{2} - 2 = -1$. Thus, the slope is steeper (more negative) for short-memory processes. The problems in this method are in principle the same as for the R/S plot and the log-log correlogram.

Figures 4.9a to 4.9d display the variance plot for several time series. It is convenient to draw a straight line with slope -1 as a reference line in the same picture. The figures illustrate that the variance plot gives us a rough idea about whether there is long memory in the data, provided that the long memory is strong enough. Slight departures from $H = \frac{1}{2}$ seem, however, rather difficult to distinguish from $H = \frac{1}{2}$, even for rather large sample sizes. Compared to the log-log correlogram, the picture may be somewhat easier to interpret.

4.5 Variogram

The variogram (or semivariogram) is often used in geostatistics. In particular, for spatial processes it is frequently used together with a methodology known under the name of kriging (see, e.g., Journel and Huijbregts 1978). The variogram at lag k is defined by

$$V(k) = \frac{1}{2}E[(X_t - X_{t-k})^2]. \qquad (4.16)$$

If X_t is stationary with covariances $\gamma(k)$ and correlations $\rho(k)$, then $V(k)$ converges to a finite value and

$$V(k) = \gamma(0)(1 - \rho(k)) = V(\infty)(1 - \rho(k)). \qquad (4.17)$$

Thus,

$$\rho(k) = 1 - \frac{V(k)}{V(\infty)}. \qquad (4.18)$$

If we replace $\rho(k)$ by the sample correlation $\hat{\rho}(k)$, then a plot of $V(k)$ against k is equivalent to the correlogram. As an alternative, a direct estimation method consists of plotting the squared differences $(X_i - X_j)^2$ against $|i - j|$ for each pair $i < j$, $(i, j = 1, ..., n)$. A smooth curve is then fitted to the scatterplot.

The distinction between short and long memory poses essentially the same problems as the correlogram. This is illustrated by Figures 4.10a to 4.10d. Long memory appears to be visible in Figure 4.10b. Due to the slow decay of the correlations, $V(k)$ converges to its asymptotic value very slowly. On the other hand, the presence of long memory is much less obvious in Figure 4.10a.

An advantage of the semivariogram as compared to the correlogram is that it is also defined for certain nonstationary processes. It therefore allows one to distingish between stationary and certain nonstationary processes. For instance, for an ARIMA(0,1,0) process $X_t = X_{t-1} + \epsilon_t$ with independent zero mean innovations ϵ_t, $V(k)$ is equal to $\frac{1}{2}k\sigma_\epsilon^2$. For a nonstationary process with linear trend $X_t = at + \epsilon_t$ we have $V(k) = \frac{1}{2}a^2k^2 + \sigma_\epsilon^2 \approx a^2k^2$.

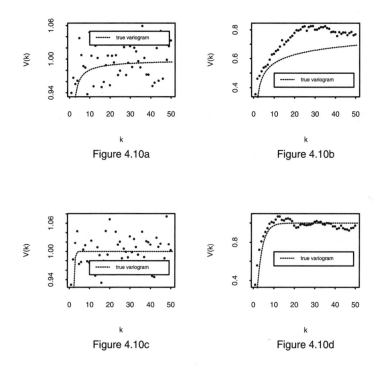

Figure 4.10a

Figure 4.10b

Figure 4.10c

Figure 4.10d

Figure 4.10. *Variogram (standardized by sample variance) for the series used in figures 4.7a,b (Figure 4.10a), figures 4.7c,d (Figure 4.10b), figures 4.8a,b (Figure 4.10c) and figures 4.8c,d (Figure 4.10d).*

4.6 Least squares regression in the spectral domain

Least squares regression in the spectral domain exploits the simple form of the pole of the spectral density at the origin:

$$f(\lambda) \sim c_f |\lambda|^{1-2H} \ (|\lambda| \to 0). \tag{4.19}$$

Equation (4.19) can be written as

$$\log f(\lambda) \sim \log c_f + (1 - 2H) \log |\lambda|. \tag{4.20}$$

Recall that, for each fixed frequency $\lambda \neq 0$, the periodogram $I(\lambda)$ is an asymptotically unbiased estimate of f; i.e., we have

$$\lim_{n \to \infty} E[I(\lambda)] = f(\lambda). \tag{4.21}$$

Usually, $I(\lambda)$ is calculated at the Fourier frequencies

$$\lambda_{k,n} = \frac{2\pi k}{n}, \ k = 1, ..., n^*, \tag{4.22}$$

where n^* is the integer part of $(n - 1)/2$. For short-memory processes, it is well known that for a finite number of frequencies $\lambda_1, ..., \lambda_k \in (0, \pi)$, the corresponding periodogram ordinates $I(\lambda_1)$, $..., I(\lambda_k)$ are approximately independent exponential random variables with means $f(\lambda_1), ..., f(\lambda_k)$. For long-memory processes, this result was given in Theorem 3.7. This, together with (4.20), leads to the approximate equation

$$\log I(\lambda_{k,n}) \approx \log c_f + (1 - 2H) \log \lambda_{k,n} + \log \xi_k, \tag{4.23}$$

where ξ_k are independent standard exponential random variables. Note that

$$E(\log \xi_k) = -C = -0.577215..., \tag{4.24}$$

where C is the Euler constant (see Gradshteyn and Ryzhik 1965, Formula 4.331). Define

$$y_k = \log I(\lambda_{k,n}), \tag{4.25}$$

$$x_k = \log \lambda_{k,n}, \tag{4.26}$$

$$\beta_o = \log c_f - C, \ \beta_1 = 1 - 2H, \tag{4.27}$$

and the "error" terms

$$e_k = \log \xi_k + C. \tag{4.28}$$

Then (4.23) can be written as

$$y_k = \beta_o + \beta_1 x_k + e_k. \tag{4.29}$$

This is a regression equation with independent identically distributed errors e_k with zero mean. The coefficients β_o and β_1 may be estimated, for instance, by least squares regression (Geweke and Porter-Hudak 1983). The estimate of H is then set equal to

$$\hat{H} = \frac{1 - \hat{\beta}_1}{2}. \tag{4.30}$$

Several problems arise with this approximate method:

1. The notion of long memory is an asymptotic one. Often the spectral density might be proportional to λ^{1-2H} in a small neighbourhood of zero only. By wrongly assuming that this proportionality is correct in the whole interval $[-\pi, \pi]$, the estimate of H can be highly biased.

2. It was shown in Theorem 3.8 that, for fixed $k \neq k'$, the periodogram at the k'th Fourier frequency is no longer asymptotically unbiased. If $\hat{\beta}_o$ and $\hat{\beta}_1$ are based on all Fourier frequencies, then this does not matter asymptotically (compare, e.g., Chapter 6). It might, however, have an influence on the finite sample properties of $\hat{H}$, or it might matter if only a small number of the smallest frequencies is used.

3. Theorem 3.8 states that, for a fixed k and k', the periodogram ordinates at λ_k and $\lambda_{k'}$ are no longer uncorrelated. As in (2), this mainly matters for the finite sample properties of $\hat{H}$, or if only a small number of the smallest frequencies is used.

4. A fourth, though less important, point is that the distribution of e_k is highly skewed. A least squares estimator will therefore be inefficient compared to an estimator that uses this property.

Point 1 can be solved, for example by estimating the least squares line from the periodogram ordinates at low frequencies only (see, e.g., Geweke and Porter-Hudak 1983, Robinson 1992a). Clearly, this can be done only at the cost of lower precision. Also, because only small frequencies are considered, problems 2 and 3 need to be taken more seriously. The fourth problem cannot be solved without abandoning the ordinary least squares method. For parametric models, efficient maximum likelihood type methods can be defined (see Chapters 5 and 6). An improved least squares method can be obtained by applying weighted least squares with appropriate weights (Robinson 1992a). Finally note that, for certain parametric classes, an efficient estimator can also be defined via generalized linear regression (see Chapter 6).

As estimation of β_o and β_1 based on all Fourier frequencies is of little practical importance (see points 1 and 4 above), we do not discuss it further here. The least squares method becomes attractive when we focus on estimating the pole (i.e., H and c_f) only, by considering a certain number of the smallest Fourier frequencies. The advantage of this method as compared to the heuristic methods described in the previous sections is that it is easier to derive the asymptotic distribution of such an estimator. In contrast to maximum likelihood estimation, almost no model assumptions are necessary. Geweke and Porter-Hudak (1983) obtained some results by heuristic arguments. A mathematical derivation of the asymptotic distribution of this semiparametric regression estimator is given in Robinson (1992a). In addition to discarding an increasing number of largest frequencies, Robinson's estimator also discards an increasing number of smallest frequencies. Thus the problems observed in Theorem 3.8 can be avoided. We state a simplified version of the result given in Robinson (1992a):

Theorem 4.6 *Let X_t be a stationary Gaussian process with long memory. Let $\hat{H}$ be the least squares estimate of H based on the Fourier frequencies λ_j, $l \leq j \leq m$. Let l and m be such that, as $n \to \infty$,*

$$m, l \to \infty, \tag{4.31}$$

$$\frac{m^5}{n^4} \to 0, \quad \frac{(\log n)^2}{m} \to 0, \tag{4.32}$$

and

$$\frac{l}{m} \to 0, \quad \frac{\sqrt{m} \log m}{l} \to 0. \tag{4.33}$$

Then, under some mild regularity conditions on f,

$$\sqrt{m}(\hat{H} - H) \to_d \sigma_o Z, \tag{4.34}$$

where Z is standard normal and

$$\sigma_o^2 = \frac{\pi^2}{24}. \tag{4.35}$$

In particular, this result means that the rate of convergence of $\hat{H}$ is slower than $n^{-\frac{1}{2}}$. We will see in Chapter 5 that, for parametric models, the rate $n^{-\frac{1}{2}}$ can be achieved. Thus, the above estimator has efficiency zero compared to an optimal parametric estimator. This is the price we pay for not assuming anything about f except the shape of the pole at zero. The efficiency loss can be rather extreme. For short series, semiparametric estimation might not be possible with reasonable accuracy. For instance, consider $m = n^{\frac{1}{2}}$. Even if l is set to zero, this would mean that for a series of 200 observations, $\hat{H}$ is estimated from 10 periodogram ordinates only. The standard error of $\hat{H}$ given by the above asymptotic result (if at all applicable for such a small value of m) is 0.203. The corresponding 95% confidence interval $\hat{H} \pm 1.96 \cdot 0.203 = \hat{H} \pm 0.397$ is almost as long as the long-memory range $\frac{1}{2} < H < 1$. Thus, it is practically impossible to distinguish long memory from short memory.

Robinson also derived the joint distribution of the scale estimate $\log \hat{c}_f = \hat{\beta}_o + C$ and $\hat{H}$ and considered the sum of peridodogram ordinates at a finite number of neighboring Fourier frequencies, instead of the raw periodogram.

An important practical problem that remains unsolved is how to choose l and m for finite samples. Depending on the choice of l and

Table 4.1. *Estimation of H by least squares regression, based on the Fourier frequencies* $\lambda_l, \lambda_{l+1}, ..., \lambda_m$.

l	m	$\hat{H}$	$m^{-\frac{1}{2}}\sigma_o$
1	144	0.527	0.053
1	100	0.591	0.064
1	60	0.712	0.083
10	144	0.446	0.053
10	100	0.525	0.064
10	60	0.786	0.083

m, results can differ considerably. Increasing m reduces the variance of $\hat{H}$ but increases the bias, unless (4.20) holds exactly. On the other hand, reducing m increases the variance but reduces the bias. A small study of the bias for finite samples can be found in Agiakloglou, Newbold, and Wohar (1993). An automatic criterion for choosing l and m may be derived, for instance, along the lines of Robinson (1991f). He considered nonparametric spectral density estimation and derived the bandwidth which minimizes the mean squared error of the estimated cumulative spectral distribution function. The optimal bandwidth turns out to depend on the unknown parameter H however.

To illustrate the edffects of l and m on $\hat{H}$, we consider the NBS weight measurements in Chapter 1. Table 4.1 gives estimates of H and the corresponding approximate standard deviation as given by Theorem 4.6 for several values of l and m. Note that the value of H increases the more one concentrates on low frequencies only. This might be an indication for long memory. However, the series appears rather short in order to make reliable inference based on the semiparametric method. More informative conclusions are obtained in Chapter 7 by a more efficient robust method.

## CHAPTER 5

Estimation of long memory: time domain MLE

5.1 Introduction

In the previous chapter, we discussed heuristic methods for estimating the long-memory parameter H. They are useful as a first informal diagnostic tool for checking if H is larger than $\frac{1}{2}$ or not. Often one is interested in more than just the question of whether there is long memory in the data. In fact, in many applications, long memory is mainly a nuisance one has to deal with rather than the actual objective of the statistical analysis. There are at least two additional types of information one might like to obtain:

1. A scale parameter. This could be, for instance, the variance of X_t or the parameter c_f in (2.2);

2. A characterization of the short-term behavior. This could be, for instance, the correlations up to a certain small lag, cyclic behavior etc.

Depending on the question one wants to answer, one needs to know H and the scale parameter c_f only, or the whole correlation structure for all lags. For instance, to obtain a confidence interval for $\mu = E(X_i)$, only H and c_f are required (see Theorem 2.2). On the other hand, to calculate an optimal prediction and a prediction interval for a future observation X_{n+k}, based on $X_1, ..., X_n$, all correlations $\rho(1), ..., \rho(n + k - 1)$ must be known or estimated. In the former case, the methods discussed in the previous chapter may be most useful. In particular, the regression method even yields approximate confidence intervals for H and c_f. These methods, however, do not tell us anything about the short-term properties of the process. More refined methods, which model the whole correlation structure, or, equivalently, the whole spectral density at all frequencies, have to be used to characterize the short-term behavior. One possible approach is to use parametric models and estimate the parameters, for example, by maximizing the likelihood. This approach is discussed in the following sections.

Apart from the necessity of modelling more than long-term properties, there are several other reasons for using parametric models and corresponding maximum likelihood type methods, instead of (or in addition to) the methods discussed in Chapter 4. Even if we need to know only the long-memory parameter H, the latter methods are less suitable for statistical inference. Correct statistical inference procedures are not known, except for the regression method. Also, the methods leave relevant questions open. For instance, the choice of the cut-off point from which the "asymptotic behavior" starts is unclear but crucial for the estimation. Moreover, in cases where we are able to build a reasonable parametric model, maximum likelihood estimation (MLE) is clearly more efficient (see Theorems 4.7 and 5.1). It should be noted, however, that from a more fundamental point of view, the question of the cut-off point is not fully solved by parametric estimation either. It is simply shifted to the problem of model choice. The question of where the "asymptotic behavior" starts is an intrinsic problem of the definition of long memory [equations (2.1), (2.2)]. Long memory is defined by an asymptotic property, namely the asymptotic decay of the correlations or the ultimate behavior of the spectral density as the frequency tends to zero. In principle, if we know H and c_f, we do not know anything about the correlations $\rho(1), ..., \rho(k_o)$ for any fixed lag k_o. In the "worst" case, it could happen that we observe $X_1, ..., X_{k_o+1}$, all correlations $\rho(1), ..., \rho(k_o)$ are zero, and $\rho(k)$ is proportional to k^{2H-2} ($1/2 < H < 1$) for $k > k_o$. Naturally, in this case, there is no way to guess that the data generating process has long memory unless we have some prior knowledge. For the estimation of parameters, such as the expected value $\mu = E(X_t)$, the slow decay of correlations beyond lag k_o would actually not matter. The distribution of statistics based on $X_1, ..., X_{k_o+1}$ depends only on the first k_o correlations. The ultimately slow decay would matter, however, for forecast intervals or if the sample size is increased. In practice, we are more likely to observe a gradual transition from an "arbitrary" correlation structure to a correlation structure of the form

$$\rho(k) = c_\rho \cdot k^{2H-2}. \tag{5.1}$$

The exact form (5.1), however, may never be achieved exactly. It is to be understood mainly as a useful simple approximation. Two approaches to solve this problem are:

1. Estimation of the whole correlation structure using a sufficiently flexible class of parametric models.
2. Semiparametric estimation, i.e., estimation of c_ρ and H, by using only correlations at large lags $k \geq k_n$. In order to ensure consistency, one chooses k_n such that, as $n \to \infty$,

$$k_n \to \infty \text{ and } k_n n^{-1} \to 0. \tag{5.2}$$

In the frequency domain this means that we estimate

$$c_f = \lim_{\lambda \to 0}[|\lambda|^{2H-1}f(\lambda)] \tag{5.3}$$

and H, by using only periodogram ordinates at small frequencies λ. In order to ensure consistency, one restricts λ to the interval $0 < \lambda < \lambda_n$ such that, as $n \to \infty$,

$$\lambda_n \to 0 \text{ and } \lambda_n n \to \infty. \tag{5.4}$$

This approach is taken by the regression method described in Section 4.6.

Approach 1 has the advantage that we are able to characterize all correlations, not just their asymptotic decay. In the spectral domain this means that we characterize the whole spectral density for all frequencies in $[-\pi, \pi]$. On the other hand, the approach is in some sense more difficult. To find a suitable parametric model that is not too complicated is not always an easy task. A poor model can lead to biased estimates of H. Analogously, choosing a wrong cut-off point leads to a biased estimate of H. A third method that uses parametric models but robust estimation in the frequency domain, is discussed in Chapter 7.

All methods considered in the following sections are based on the Gaussian likelihood function. This does not mean that they can be used only for Gaussian time series. At least for some of the methods, the same central limit theorems are known to hold for more general processes. Some caution is needed, however, when applying these methods to non-Gaussian processes (see Chapter 7). The main reason why one first concentrates on Gaussian MLE is simplicity. A Gaussian process is fully characterized by its mean and its covariances (or the spectral density). MLEs can therefore be expressed in terms of the first two moments only.

5.2 Some definitions and useful results

Suppose that $X_1, X_2, ..., X_n$ is a stationary process with mean μ and variance σ^2. We assume that (2.1) and (2.2) hold with $1/2 <$

$H < 1$. Let the spectral density f be characterized by an unknown finite dimensional parameter vector

$$\theta^o = (\sigma_o^2, H^o, \theta_3^o, ..., \theta_M^o). \qquad (5.5)$$

Thus, we assume that the spectral density comes from a parametric family of densities $f(\lambda) = f(\lambda; \theta)$, where $\theta \in \Theta \subset R^M$. Given the observations $X_1, ..., X_n$, one would like to estimate the unknown parameter vector θ^o. We will use the following notation:

$$X = (X_1, X_2, ..., X_n)^t,$$

$$\Sigma_n(\theta) = [\gamma(j - l)]_{j,l=1,...,n} = \text{covariance matrix of } X,$$

$$|\Sigma_n| = \text{determinant of } \Sigma_n.$$

Without loss of generality, we will assume μ to be known and equal to zero. This is in particular justified by the fact that the asymptotic distribution of the Gaussian maximum likelihood estimator (MLE) of θ turns out to be the same whether μ is replaced by the sample mean or not. Furthermore, X_t will be assumed to be a causal invertible linear process. That is, X_t can be expressed in two ways:

$$X_t = \sum_{s=1}^{\infty} b(s)X_{t-s} + \epsilon_t \qquad (5.6)$$

and

$$X_t = \sum_{s=0}^{\infty} a(s)\epsilon_{t-s} \qquad (5.7)$$

where the innovations ϵ_s are uncorrelated random variables with zero mean and variance $\text{var}(\epsilon_s) = \sigma_\epsilon^2$. Conditions (5.6) and (5.7) imply that a new observation X_t depends only on the past, and it depends on the past in a linear way. The former (causality) is a natural assumption in most applications. The latter (linearity), is mathematically convenient, but does not need to hold in practise. For time series where X_t depends on the past in a nonlinear way see, e.g., Tong (1993). Here we restrict attention to linear processes.

Before discussing several estimation methods based on the Gaussian likelihood function, we note some useful results. The first result gives the asymptotic behavior of the coefficients $b(s)$ and $a(s)$.

Lemma 5.1 *There exist constants $0 < c_b < \infty$ and $0 < c_a < \infty$ such that, as $k \to \infty$,*

$$b(s) \sim c_b k^{-H-\frac{1}{2}} \qquad (5.8)$$

and

$$a(s) \sim c_a k^{H-\frac{3}{2}}.$$ (5.9)

In particular, we have

$$\sum_{s=1}^{\infty} b(s) < 0$$ (5.10)

and

$$\sum_{s=1}^{\infty} a(s) = \infty.$$ (5.11)

The second useful result refers to the connection between one-step ahead prediction and the choice of the scale parameter. Suppose we want to predict a future value of X_t from the infinite past $X_s, s \le t - 1$. Denote by $\hat{X}_t$ the prediction of X_t. Among all linear predictions, the mean squared prediction error

$$MSPE = E[(X_t - \hat{X}_t)^2 | X_s, s \le t - 1]$$ (5.12)

is minimized by

$$\hat{X}_t = E[X_t | X_s, s \le t - 1] = \sum_{s=1}^{\infty} b(s) X_{t-s}.$$ (5.13)

The minimal prediction error is given by (see, e.g., Priestley 1981, p. 741):

Lemma 5.2

$$MSPE = \sigma_\epsilon^2 = 2\pi \exp[\frac{1}{2\pi} \int_{-\pi}^{\pi} \log f(\lambda; \theta^o) d\lambda].$$ (5.14)

In particular it follows that, for

$$f_1(\lambda) = \frac{2\pi}{\sigma_\epsilon^2} f(\lambda),$$ (5.15)

we have

$$\int_{-\pi}^{\pi} \log f_1(\lambda) d\lambda = 0.$$ (5.16)

5.3 Exact Gaussian MLE

Suppose that, in addition to (5.6) and (5.7), X_t is a Gaussian process. Then the joint distribution function of $X = (X_1, X_2, ..., X_n)^t$ is equal to

$$h(x; \theta^o) = (2\pi)^{-\frac{n}{2}} |\Sigma(\theta^o)|^{-\frac{1}{2}} e^{-\frac{1}{2} x^t \Sigma^{-1}(\theta^o) x}.$$ (5.17)

Here, $x = (x_1, ..., x_n)^t \in R^n$. The log-likelihood function is given by

$$L_n(x; \theta^o) = \log h(x; \theta^o)$$
$$= -\frac{n}{2} \log 2\pi - \frac{1}{2} \log |\Sigma(\theta^o)| - \frac{1}{2} x^t \Sigma^{-1}(\theta^o) x. \quad (5.18)$$

Define the M-dimensional vector

$$L'_n(x; \theta) = \frac{\partial}{\partial \theta_j} L_n(x; \theta)$$

$$= -\frac{1}{2} \frac{\partial}{\partial \theta_j} \log |\Sigma(\theta)| - \frac{1}{2} x^t [\frac{\partial}{\partial \theta_j} \Sigma^{-1}(\theta)] x \quad (j = 1, 2, ..., M). \quad (5.19)$$

The MLE of θ^o is obtained by maximizing $\log h(x; \theta)$ with respect to the M-dimensional parameter vector θ. Under mild regularity assumptions, this maximization problem can be reformulated in terms of the first partial derivatives. The MLE $\hat{\theta}$ is the solution of the system of M equations

$$L'_n(x; \hat{\theta}) = 0. \quad (5.20)$$

The asymptotic distribution of $\hat{\theta}$ can be derived by looking at the Taylor expansion of L'_n. Denote by

$$L''_n(x; \theta) = \frac{\partial^2}{\partial \theta_j \theta_l} L_n(x; \theta) \quad (j, l = 1, ..., M) \quad (5.21)$$

the matrix of second partial derivatives of L_n. From (5.20) we obtain

$$L'_n(x; \hat{\theta}) = 0 = L'_n(x; \theta^o) + L''_n(x; \theta^o)(\hat{\theta} - \theta^o) + r_n. \quad (5.22)$$

If one can show that r_n is asymptotically negligible, then the asymptotic distribution of $\hat{\theta} - \theta^o$ is equal to the asymptotic distribution of

$$- [L''_n(x; \theta^o)]^{-1} L'_n(x; \theta^o). \quad (5.23)$$

Yajima (1985) and Dahlhaus (1989) prove (5.23) and the following limit theorem. Dahlhaus derives the result for general Gaussian processes. Yajima assumes normal and non-normal fractional ARIMA$(0,d,0)$ processes.

Theorem 5.1 Let X_t be a Gaussian process with the properties given in Section 5.2 and let $\hat{\theta}$ be defined by (5.20). Define the $M \times M$-matrix $D = [D_{ij}]_{i,j=1,...,M}$ by

$$D_{ij}(\theta^o) = \frac{1}{2\pi} \int_{-\pi}^{\pi} \frac{\partial}{\partial \theta_i} \log f(x) \frac{\partial}{\partial \theta_j} \log f(x) dx|_{\theta=\theta^o} \quad (5.24)$$

and

$$C(\theta^o) = 2D^{-1}(\theta^o).\qquad(5.25)$$

Under a few additional regularity conditions on f, the following holds as $n \to \infty$:

(i)

$$\hat{\theta} \to \theta^o\qquad(5.26)$$

almost surely, and

(ii)

$$n^{\frac{1}{2}}(\hat{\theta} - \theta^o) \to_d \zeta.\qquad(5.27)$$

where ζ is an M-dimensional normal random vector with zero mean and covariance matrix C.

In particular, the result implies that $\hat{\theta}$ has the same rate of convergence $n^{-\frac{1}{2}}$ as corresponding parameter estimates for short-memory processes. This is in contrast to location estimators that have a slower rate of convergence, if long-range dependence is present (see section 1.1 and Chapter 8).

The covariance matrix C is diagonal if the functions

$$f_j = \frac{\partial}{\partial\theta_j}\log f \ (j = 1, ..., M)$$

are orthogonal in the sense that

$$\int_{-\pi}^{\pi} f_j(\lambda)f_l(\lambda)d\lambda = 0 \ (j \neq l).$$

If in addition, f_j does not depend on the unknown θ, then the variance of $\hat{\theta}_j$ is the same for all θ. Consider, for example, a fractional ARIMA$(0, d, 0)$ process. The spectral density is equal to

$$\frac{\sigma_\epsilon^2}{2\pi}|1 - e^{i\lambda}|^{1-2H} = \frac{\sigma_\epsilon^2}{2\pi}(2 - 2\cos\lambda)^{\frac{1}{2}-H}.$$

Define the parameter vector

$$\theta = (\frac{\sigma_\epsilon^2}{2\pi}, H).$$

From Gradshteyn and Rhyzik (1965, Formulas 4.224 No. 9 and 7 respectively) we have

$$\int_{-\pi}^{\pi} \log(2 - 2\cos\lambda)d\lambda = 0$$

and

$$\int_{-\pi}^{\pi} [\log(2 - 2\cos\lambda)]^2 d\lambda = \frac{2}{3}\pi^3.$$

Therefore, $\hat{H}$ is asymptotically independent of $\hat{\sigma}_\epsilon^2 = 2\pi\hat{\theta}_1$ with asymptotic variance

$$\frac{4\pi}{\int_{-\pi}^{\pi}[\log(2 - 2\cos\lambda)]^2 d\lambda} = \frac{6}{\pi^2} \approx 0.6079.$$

The estimator of $\hat{\sigma}_\epsilon^2$ has the asymptotic variance $2\sigma_\epsilon^4$.

It is well known that, under suitable regularity conditions, the MLE is asymptotically efficient in the sense of Fisher. This means that the Fisher information matrix

$$\Gamma_n(\theta_o) = E\{[L_n'(x;\hat{\theta})][L_n'(x;\hat{\theta}]^t\} \tag{5.28}$$

converges to the inverse of the asymptotic covariance matrix of $\hat{\theta}$. Recall that the inverse of

$$\Gamma(\theta^o) = \lim_{n\to\infty} \Gamma_n(\theta^o) \tag{5.29}$$

is called the asymptotic Cramer-Rao bound. One of the usual regularity conditions for establishing efficiency of the MLE is that the spectral density is continuous everywhere in the interval $[-\pi, \pi]$. In the case of long memory, this condition does not hold. A new proof is needed to establish efficiency of the MLE. This was done by Dahlhaus (1989). He showed that, for Gaussian processes with long memory, the Cramer-Rao bound is achieved by the MLE:

Theorem 5.2 *Under the assumptions of Theorem 5.1,*

$$\lim_{n\to\infty} \Gamma_n(\theta^o) = C^{-1}(\theta^o). \tag{5.30}$$

We conclude this section by discussing briefly a computational aspect. The calculation of the log-likelihood function L_n or its derivative requires the inversion of the covariance matrix Σ_n. For computational reasons, one might want to avoid this inversion (see the discussion in the next section). An alternative to calculating L_n via (5.18) directly isscale large bnewnew.asc to decompose the $n$dimensional Gaussian likelihood function into a product of one-dimensional conditional distributions. The distribution $h(x)$ can be written as

$$h(x) = h_1(x_1)h_2(x_2|x_1)\cdots h_n(x_n|x_1, x_2, ..., x_n). \tag{5.31}$$

Here, $h_j(x_j|x_1, ..., x_{j-1})$ denotes the conditional distribution of X_j given $X_1 = x_1, ..., X_{j-1} = x_{j-1}$. Because the probability densities $h_1, ..., h_n$ are one-dimensional normal distribution functions, they are fully determined by their mean and variance. The mean and variance of h_j are denoted by μ_j and σ_j^2. The mean μ_j is equal to the best linear prediction of X_j given $X_1, ..., X_{j-1}$,

$$\mu_j = E(X_j|X_1, ..., X_{j-1}) = \hat{X}_j = \sum_{s=1}^{j-1} \beta_{j-1,s} X_{j-s}. \qquad (5.32)$$

The coefficients $\beta_{j-1,s}$ are the partial correlations. The variance of h_j is equal to the expected mean squared error of $\hat{X}_j$,

$$\sigma_j^2 = E[(\hat{X}_j - \mu_j)^2|X_1, ..., X_{j-1}]. \qquad (5.33)$$

The coefficients $\beta_{j,s}$ are given by (see, e.g., Durbin 1960)

$$\Omega^{(j)}\beta^{(j)} = \rho^{(j)}. \qquad (5.34)$$

Here,

$$\rho^{(j)} = (\rho(1), ..., \rho(j))^t,$$
$$\beta^{(j)} = (\beta_{j,1}, \beta_{j,2}..., \beta_{j,j})^t$$

and

$$\Omega^{(j)} = [\rho(r - s)]_{r,s=1,...,j}$$

is the correlation matrix of $X_1, ..., X_{j+1}$. Equation (5.34) can be solved by inverting the correlation matrix $\Omega^{(j)}$:

$$\beta^{(j)} = [\Omega^{(j)}]^{-1}\rho^{(j)}. \qquad (5.35)$$

A more elegant and computationally faster solution is the Durbin-Levinson algorithm (Durbin 1960, see also, e.g., Brockwell and Davis 1987, pp.162-163).

5.4 Why do we need approximate MLE's?

At first sight, the problem of estimating θ^o by the maximum likelihood method seems to be solved by Theorem 5.1. However, the exact MLE poses computational problems. To obtain the solution of (5.20), (5.19) has to be evaluated for many trial values of θ. This can be costly in terms of CPU time, in particular if the dimension of θ is high or if we have a long time series. Also, for long time series, storing the whole covariance matrix $\Sigma(\theta)$ in its raw form requires excessive computer memory. Finally, evaluation of the inverse of the covariance matrix may be numerically unstable. In particular, if H is close to 1, the covariances change very slowly because of their slow decay to zero, so that $\Sigma(\theta)$ can become almost singular. For instance, for $n = 100$, the determinant of the correlation matrix of fractional Gaussian noise with $H = 0.5, 0.6, 0.7, 0.8$, and 0.9 is equal to $1, 0.06, 2.2 \cdot 10^{-6}, 2.7 \cdot 10^{-16}$, and $5.0 \cdot 10^{-39}$, respectively (the numbers are rounded). Also, the ratio of the largest eigenvalue divided by the smallest eigenvalue is approximately equal to $1, 3, 11, 43$, and 222, respectively. The problem of inverting Σ_n can be avoided by the recursive calculation of the normal density function h via (5.31). However, this recursive algorithm needs to be run for each trial value of θ. This may lead to long CPU times, in particular for long time series and/or for high dimensional parameter vectors θ. For instance, for a series of length $n = 6574$, Haslett and Raftery (1989) reported a CPU time of about 45 hours on a VAX11/780 for the calculation of the exact MLE, based on this algorithm. An alternative to solving the exact maximum likelihood equations is to maximize an approximation to the likelihood function. In the following sections, we discuss several possible approaches to approximating the Gaussian likelihood function.

5.5 Whittle's approximate MLE

The two terms in (5.18) that depend on θ are the logarithm of the determinant of the covariance matrix,

$$\log |\Sigma_n(\theta)| \tag{5.36}$$

and the quadratic form

$$x^t \Sigma^{-1}(\theta) x. \tag{5.37}$$

Let us look at each of these terms separately.

1. Approximation of $\log |\Sigma_n(\theta)|$: It was shown by Grenander and Szegö (1958) that

$$\lim_{n \to \infty} \frac{1}{n} \log |\Sigma_n(\theta)| = \frac{1}{2\pi} \int_{-\pi}^{\pi} \log f(\lambda; \theta) d\lambda. \tag{5.38}$$

Therefore, we replace $\log |\Sigma_n(\theta)|$ by $n(2\pi)^{-1} \int_{-\pi}^{\pi} \log f(\lambda; \theta) d\lambda$.

2. Approximation of $x^t \Sigma^{-1}(\theta) x$: The matrix $\Sigma^{-1}(\theta)$ can be replaced by a matrix whose elements are easier to calculate. Define

$$A(\theta) = [\alpha(j - l)]_{j,l=1,\dots,n} \tag{5.39}$$

to be the $n \times n$-matrix with elements

$$\alpha(j - l) = (2\pi)^{-2} \int_{-\pi}^{\pi} \frac{1}{f(\lambda; \theta)} e^{i(j-l)\lambda} d\lambda. \tag{5.40}$$

The matrix A is asymptotically the inverse of Σ_n in the following sense:

Lemma 5.3 *Let*

$$A_\infty(j,l) = \alpha(j-l), \quad j,l = 1,2,\dots, \tag{5.41}$$

$$\Sigma_\infty(j,l) = \gamma(j-l), \quad j,l = 1,2,\dots \tag{5.42}$$

and define the infinite dimensional identity matrix I_∞ by

$$I_\infty(j,l) = \delta(j-l), \tag{5.43}$$

where $\delta(s) = 1$ for $s = 0$ and $\delta(s) = 0$ otherwise. Also, define the products $\Sigma_\infty A_\infty$ and $A_\infty \Sigma_\infty$ by

$$[\Sigma_\infty A_\infty](j,l) = \sum_{k=-\infty}^{\infty} \Sigma_\infty(k,j)A_\infty(k,l) \tag{5.44}$$

and

$$[A_\infty \Sigma_\infty](j,l) = \sum_{k=-\infty}^{\infty} A_\infty(k,j)\Sigma_\infty(k,l). \tag{5.45}$$

Then A_∞ is the inverse of Σ_∞ in the sense that

$$A_\infty \Sigma_\infty = \Sigma_\infty A_\infty = I_\infty. \tag{5.46}$$

Another interpretation of A can be given in terms of the best linear prediction of X_t given the infinite past and the infinite future (see, e.g., Künsch 1981):

Lemma 5.4 *Consider the prediction of X_t, given all past observations X_s, $s \leq t-1$, and all future observations X_s, $s \geq t+1$. Among all linear predictors, the mean squared prediction error $MSPE = E[(X_t - \hat{X}_t)^2 | X_s, s \neq t]$ is minimized by*

$$\hat{X}_t = \alpha(o)^{-1} \sum_{s \in Z, s \neq 0} \alpha(s)X_{t+s}. \tag{5.47}$$

Combining approximations 1 and 2, (5.22) is approximated by

$$L_n^* = -\frac{n}{2}\log 2\pi - \frac{n}{2}\frac{1}{2\pi}\int_{-\pi}^{\pi} \log f(\lambda;\theta)d\lambda - \frac{1}{2}x^t A(\theta)x. \tag{5.48}$$

Only the last two terms depend on the parameter vector θ. An approximate MLE of θ^o is obtained by minimizing the function

$$L_W(\theta) = \frac{1}{2\pi}\int_{-\pi}^{\pi} \log f(\lambda;\theta)d\lambda + \frac{x^t A(\theta)x}{n} \tag{5.49}$$

with respect to θ. The subscript W stands for Whittle, who proposed this approximation in the context of short-memory time series (Whittle 1953). Under mild regularity conditions, minimizing

(5.49) is equivalent to solving the system of nonlinear equations

$$\frac{\partial}{\partial \theta_j} L_W(\theta)|_{\theta=\hat{\theta}} = 0, \qquad (5.50)$$

with $j = 1, 2, ..., M$. Written explicitly, (5.50) is of the form

$$\frac{\partial}{\partial \theta_j} \frac{1}{2\pi} \int_{-\pi}^{\pi} \log f(\lambda; \theta) d\lambda + \frac{1}{n} \frac{\partial}{\partial \theta_j} x^t A(\theta) x = 0. \qquad (5.51)$$

A simplification of (5.50) can be achieved by choosing a special scale parameter. Define

$$\eta = (\theta_2, \theta_3, ..., \theta_M) \qquad (5.52)$$

and

$$\theta^* = (1, \eta). \qquad (5.53)$$

Choose the scale parameter θ_1 such that

$$f(\lambda; \theta) = \theta_1 f(\lambda; \theta^*) \qquad (5.54)$$

and

$$\int_{-\pi}^{\pi} \log f(\lambda; \theta^*) d\lambda = 0. \qquad (5.55)$$

Thus, $\theta_1 = 2\pi\sigma_\epsilon^2$ where σ_ϵ^2 is the optimal one-step-ahead prediction error (see Lemma 5.2). We will use the following notation:

$$\dot{A}_j(\theta) = [\dot{\alpha}_j(r - s; \theta)]_{r,s=1,...,n} = \frac{\partial}{\partial \theta_j} A(\theta) \quad (j = 1, ..., M),$$

$$Q(\theta) = X^t A(\theta) X = \sum_{r,s=1}^{n} \alpha(r - s; \theta) X_r X_s,$$

$$\dot{Q}_j(\theta) = x^t \dot{A}_j(\theta) x \quad (j = 1, 2, 3, ...)$$

and

$$\dot{Q}(\theta) = (\dot{Q}_1(\theta), ..., \dot{Q}_M(\theta))^t.$$

Also, $\ddot{Q}$ will denote the $M \times M$-matrix of all second partial derivatives of Q. Finally, we set

$$Q^*(\eta) = Q(\theta^*)$$

with corresponding first and second derivatives with respect to η, $\dot{Q}^*(\eta)$ and $\ddot{Q}^*(\eta)$. Due to (5.55), the first term in (5.51) is equal to zero. Equation (5.51) therefore simplifies to the following steps:

1. Obtain $\hat{\eta}$ by solving

$$\dot{Q}^*(\hat{\eta}) = 0. \qquad (5.56)$$

2. Set $\hat{\sigma}_\epsilon^2 = 2\pi\hat{\theta}_1$ equal to

$$\hat{\sigma}_\epsilon^2 = 2\pi Q^*(\hat{\eta}). \tag{5.57}$$

The same simplification could be achieved by setting the integral of $\log f$ equal to any arbitrary constant, not necessarily zero. However, letting the constant zero seems most natural, because then the scale parameter θ_1 has a nice intuitive interpretation.

As in the previous chapter, a Taylor expansion can be applied to obtain the asymptotic distribution of $\hat{\theta}$. First note that, (5.56) and (5.57) can also be written as the system of M equations

$$\dot{Q}(\hat{\theta}) - E[\dot{Q}] = 0. \tag{5.58}$$

Because the components of $\dot{Q}$ are quadratic forms, the limiting distribution of $\hat{\theta}$ follows by Taylor expansion from the results in Chapter 3. The following central limit theorem is a summary of results by Fox and Taqqu (1986), Beran (1986), and Dahlhaus (1989).

Theorem 5.3 *Let $\hat{\theta}$ be defined by (5.56) and (5.57). Then, under the conditions of Theorem 5.1,*

$$\hat{\theta} \rightarrow \theta^o \tag{5.59}$$

almost surely, and

$$\sqrt{n}(\hat{\theta} - \theta^o) \rightarrow_d \zeta, \tag{5.60}$$

where ζ is defined in Theorem 5.1.

Thus, Whittle's approximate MLE has the same asymptotic distribution as the exact MLE. It is therefore asymptotically efficient for Gaussian processes. It should be noted that (5.55) implies not only a simplification of (5.51), but also the asymptotic covariance matrix of $\hat{\theta}$ has a simpler form. From (5.25) we obtain $D_{1j} = D_{j1} = 0$, $(j \neq 1)$ and thus $\text{cov}(\hat{\theta}_1, \hat{\theta}_j) = 0$ $(j \neq 1)$. Thus, the scale estimate (5.57) is asymptotically independent of the other parameter estimates. This is a nice property, because σ_ϵ^2 is usually a nuisance parameter only.

In analogy to the corresponding central limit theorem for quadratic forms, Theorem 5.3 can be generalized to linear (not necessarily Gaussian) processes, at least for $\hat{\eta}$ (Giraitis and Surgailis 1990a).

Theorem 5.4 *If X_t is a linear process with long memory defined by (3.12), then (under a few regularity conditions), the result of Theorem 5.3 follows for $\hat{\eta}$ (with θ^o replaced by η^o).*

A remark should be made about computational aspects. Approximation (5.51) replaces the inversion of the covariance matrix by the calculation of n Fourier coefficients (5.40) This calculation needs to be repeated for each trial value of θ. For models where no explicit formula for (5.40) is available, numerical integration is necessary. Depending on the efficiency of the program that performs this integration, and depending on the complexity of the function f^{-1}, repeated calculation of (5.40) may still require large CPU times. In first approximation, one may therefore consider replacing the integral in (5.40) by a suitable Riemann sum. This leads to a fast algorithm. It is discussed in more detail in Section 6.1.

5.6 An approximate MLE based on the AR representation

5.6.1 Definition for stationary processes

The idea of Whittle's approximate MLE is to replace the elements of the inverse covariance matrix Σ_n^{-1} by the corresponding elements of the matrix A, which is a part of the two-sided infinite dimensional matrix A_∞. As we saw in Lemma 5.4, A_∞ also gives rise to the best linear prediction of X_t, given all past and all future observations. An alternative approximate MLE can be obtained by using the best linear prediction of X_t given all past values $X_{t-1}, X_{t-2}, \dots$ only. This prediction follows directly from the infinite autoregressive representation (5.6) of X_t. If we knew the infinite past X_s $(s \leq t)$, then we could reconstruct the sequence of independent identically distributed innovations ϵ_s $(s \leq t)$ by

$$\epsilon_t = X_t - \sum_{s=1}^{\infty} b(s)X_{t-s}. \tag{5.61}$$

If X_t is Gaussian, then the log-likelihood function of $\epsilon_1, \dots, \epsilon_n$ is given by

$$L_n = -\frac{n}{2}\log 2\pi - \frac{n}{2}\log \sigma_\epsilon^2 - \frac{1}{2}\sum_{t=1}^{n}(\frac{\epsilon_t}{\sigma_\epsilon})^2. \tag{5.62}$$

If instead of the infinite past, only a finite number of past values is observed, the innovations $\epsilon_2, \dots, \epsilon_n$ can be estimated by

$$u_t = X_t - \sum_{s=1}^{t-1} b(s)X_{t-s}, \quad (t = 2, \dots, n). \tag{5.63}$$

This amounts to assuming that $X_t = 0$ for $t \leq 0$. An approximate log-likelihood function can be defined by replacing ϵ_t in (5.62) by u_t. For a parametric model, the coefficients $b(s)$ depend on a parameter vector η^o. An approximate MLE of $\theta^o = (\sigma_\epsilon^2, \eta)$ is obtained by maximizing the approximate log-likelihood function with respect to θ. This can be written as follows. Define

$$r_t(\theta) = \frac{u_t(\eta)}{\sqrt{\theta_1}}, \tag{5.64}$$

$$\dot{r}_t(\theta) = (\frac{\partial}{\partial \theta_1} r_t(\theta), ..., \frac{\partial}{\partial \theta_M} r_t(\theta))^t, \tag{5.65}$$

and

$$\dot{u}_t(\eta) = (\frac{\partial}{\partial \eta_1} u_t(\eta), ..., \frac{\partial}{\partial \eta_{M-1}} u_t(\eta))^t. \tag{5.66}$$

The approximate MLE $\hat{\theta}$ is obtained by minimizing the sum of the function

$$n \log \theta_1 + \sum_{t=2}^{n} r_t^2(\theta) \tag{5.67}$$

with respect to θ. Taking the first partial derivatives with respect to θ leads approximately to the system of M nonlinear equations

$$\sum_{t=2}^{n} \{r_t(\theta)\dot{r}_t(\theta) - E[r_t(\theta)\dot{r}_t(\theta)]\} = 0. \tag{5.68}$$

Because of the special role of the scale parameter, this can be split into two parts,

$$\sum_{t=2}^{n} \dot{u}_t(\eta)u_t(\eta) = 0 \tag{5.69}$$

and

$$\sigma_\epsilon^2 = \frac{1}{n-1} \sum_{t=2}^{n} u_t^2(\theta). \tag{5.70}$$

The same central limit theorem as for Whittle's estimator holds (Beran 1993b):

Theorem 5.5 *Suppose that the assumpions of Theorem 5.3 hold. Let $\hat{\theta}$ be the solution of (5.68). Then the conclusion of Theorem 5.3 follows.*

Equation (5.68) is analogous to the approximate maximum likelihood equations for finite-order autoregressive processes and to the normal equations in regression. The difference is that there is no finite lag s_o such that $b(s)$ is equal to zero for all $s > s_o$. Therefore, we cannot reconstruct any of the innovations ϵ_t exactly from the observations $X_1, ..., X_n$, even if we know the exact coefficients $b(s), (s = 1, 2, ...)$. In particular, the variance of the approximate residuals u_t depends on t instead of being constant. Asymptotically, this effect is negligible. For small sample sizes, it might be worthwhile to replace the coefficients $b(s)$ by the coefficients that give the best linear prediction of X_t given the finite past $X_{t-1}, ..., X_1$. These coefficients and the variance of the resulting residuals can be calculated recursively by the Durbin-Levinson algorithm (see also Haslett and Raftery 1989).

5.6.2 Generalization to nonstationary processes; a unified approach to Box-Jenkins modelling

To conclude this section, a few remarks on estimation for nonstationary ARIMA(p, d, q) models should be made. It was noted in Beran (1994) that the infinite AR representation is not restricted to the case where X_t is stationary (i.e. $d < 1/2$), but can be extended to any $d > -1/2$. The approximate MLE (5.68) is therefore defined for any, stationary or nonstationary, fractional or nonfractional, invertible ARIMA(p, d, q) process. In particular, this leads to a unified approach to Box-Jenkins time series modelling. Recall that in traditional time series modelling with ARIMA(p, d, q) models, d is assumed to be an integer. "Estimation" of d is usually done by "trial and error" accompanied by some diagnostic plots. Once d is found, the fact that d had to be estimated is ignored. Confidence intervals for the parameters are calculated as if d had been known a priori. In contrast, the method in Beran (1994) estimates d (and all other parameters) by maximizing the likelihood. Confidence intervals for the parameters take into account that d is estimated. For a detailed discussion of asymptotic results, simulations, data examples and algorithms see Beran (1994). For related results in the frequency domain see also, e.g., Hurvich and Ray (1994).

CHAPTER 6

Estimation of long memory: frequency domain MLE

6.1 A discrete version of Whittle's estimator

Whittle's approximate maximum likelihood estimator (MLE) requires the calculation of n integrals (5.44) for each trial value of θ. This can be a time consuming task, in particular for large sample sizes and if θ has a high dimension. Note however that, in contrast to the spectral density f itself, the function $1/f$ is "well behaved" at the origin. A simple Riemann sum

$$\tilde{\alpha}(k) = 2 \frac{1}{(2\pi)^2} \sum_{j=1}^{m} \frac{1}{f(\lambda_{j,m})} e^{ik\lambda_{j,m}} \frac{2\pi}{m} \qquad (6.1)$$

with

$$\lambda_{j,m} = \frac{2\pi j}{m} \quad (j = 1, ..., m^*) \qquad (6.2)$$

approximates the integral (5.40) reasonably well. Here, m^* denotes the integer part of $(m-1)/2$. The following estimator is then obtained: Note first that (5.49) can be written in terms of the periodogram $I(\lambda)$ as

$$L_W(\theta) = \frac{1}{2\pi} [\int_{-\pi}^{\pi} \log f(\lambda; \theta) d\lambda + \int_{-\pi}^{\pi} \frac{I(\lambda)}{f(\lambda)} d\lambda]. \qquad (6.3)$$

From (6.2) one obtains the approximation

$$\tilde{L}_W(\theta) = 2 \frac{1}{2\pi} [\sum_{j=1}^{m^*} \log(\lambda_{j,m}; \theta) \frac{2\pi}{m} + \sum_{j=1}^{m^*} \frac{I(\lambda_{j,m})}{f(\lambda_{j,m}; \theta)} \frac{2\pi}{m}]. \qquad (6.4)$$

Because the periodogram can be calculated by the fast Fourier transform, $\tilde{L}_W$ can be calculated very fast. Moreover, if $\theta = (\theta_1, \eta)$ is such that (5.54) and (5.55) hold, the first term in (6.3) can be replaced by $\log \theta_1$. Minimizing (6.3) amounts to minimizing the sum of the ratios $I(\lambda_{j,m}) f^{-1}(\lambda_{j,m}; \theta^*)$ with respect to η, where

$\theta^* = (1, \eta)$, and setting $\hat{\sigma}_\epsilon^2 = 2\pi\hat{\theta}^*{}_1$ equal to this sum multiplied by $4\pi/m$. In particular, if we choose $m = n$, then $\lambda_{j,n}$ are the Fourier frequencies and $\hat{\theta}$ is defined by the following two steps:

1. Minimize

$$\tilde{Q}(\eta) = \sum_{j=1}^{m^*} \frac{I(\lambda_{j,n})}{f(\lambda_{j,n}; \theta^*)} \tag{6.5}$$

with respect to η.

2. Set

$$\hat{\sigma}_\epsilon^2 = 2\pi\hat{\theta}_1 = \frac{4\pi}{n}\tilde{Q}(\hat{\eta}). \tag{6.6}$$

This approximation was suggested by Graf (1983) for fractional Gaussian noise. He derived it in a different way, by assuming that periodogram ordinates at distinct Fourier frequencies are approximately independent exponential random variables with expected value equal to the spectral density. Strictly speaking, this derivation is not quite correct though, as consistency and independence are proven for a finite number of non-zero frequencies only (cf. Theorems 3.7 and 3.8). By heuristic arguments and simulations, Graf (1983) demonstrated that for fractional Gaussian noise, limit Theorem 5.3 holds for $\hat{\theta}$ defined by (6.5) and (6.6). This is likely to hold for any linear process for which certain regularity conditions (such as those in Theorem 5.4) hold. A related consistency result is given in Theorem 3.9.

Examples:

1. We fit two models, fractional Gaussian noise and a fractional ARIMA(0,d,0) model, to the Nile River data, after subtracting the sample mean. First note that for fractional Gaussian noise with $\theta = (\sigma_\epsilon^2/(2\pi), H)$, the spectral density can be written as

$$f(\lambda; \theta) = \frac{\sigma_\epsilon^2}{2\pi}[c(H)f_1(\lambda; H)] \tag{6.7}$$

where

$$\begin{aligned} f_1(\lambda; H) = \quad & \sigma^2\pi^{-1}\Gamma(2H + 1)\sin(\pi H)(1 - \cos\lambda) \\ & \cdot \sum_{j=-\infty}^{\infty}|2\pi j + \lambda|^{-2H-1}, \end{aligned} \tag{6.8}$$

$\sigma^2 = \mathrm{var}(X_t)$ and

$$c(H) = \exp[-(2\pi)^{-1}\int_{-\pi}^{\pi}\log f_1(\lambda)d\lambda]. \tag{6.9}$$

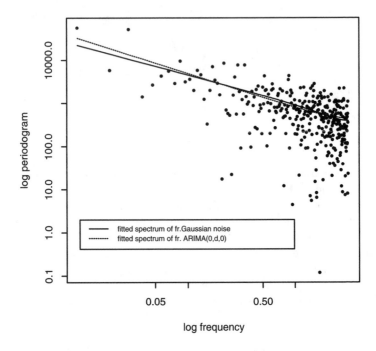

Figure 6.1. *Nile River minima: periodogram and fitted spectral densities (log-log coordinates).*

Then (5.55) holds for f_1. For a fractional ARIMA$(0, d, 0)$ process,

$$f(\lambda; \theta) = \frac{\sigma_\epsilon^2}{2\pi} |1 - e^{i\lambda}|^{1-2H} \tag{6.10}$$

and (5.55) holds for

$$f_1(\lambda; H) = |1 - e^{i\lambda}|^{1-2H}. \tag{6.11}$$

The estimates of H are equal to 0.84 and 0.90, respectively. The 95% confidence intervals for H are $[0.79, 0.89]$ and $[0.84, 0.96]$, respectively. The plot of the periodogram and the fitted spectral densities shows a good agreement between data and both fitted models (Figure 6.1).

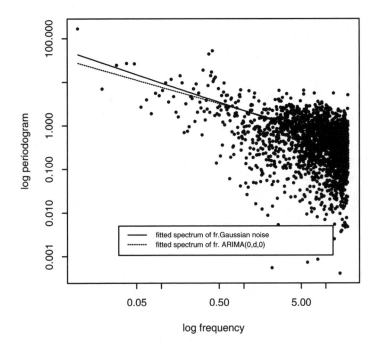

Figure 6.2. *Ethernet data: periodogram and fitted spectral densities (log-log coordinates).*

2. The same models are fitted to the Ethernet data. The estimated values of $\hat{H}$ are 0.82 and 0.77, with 95% confidence intervals $[0.79, 0.84]$ and $[0.75, 0.79]$, respectively. Both fitted spectral
densities agree well with the observed periodogram. It should be
noted, however, that the spectral density characterizes a process
fully (together with the mean) only if the process is Gaussian.
While this might be the case in good approximation for the
Nile River data (compare Figure 7.1), the Ethernet data are
clearly non-Gaussian (see Figure 1.7). A more refined analysis
that takes into account the discrete nature of the observations
would be needed.

6.2 Estimation by generalized linear models

In general, the approximate method in the previous section is numerically simple, if the only two parameters to be estimated are σ_ϵ^2 and $\eta = H$. For higher-dimensional parameter vectors η, (6.5) is a system of nonlinear equations. Depending on the complexity of the model, finding the solution of (6.5) can be numerically difficult and might require long CPU times. This motivates the search for models where this task can be simplified. A class of models for which (6.5) reduces to the estimation of the parameters in a generalized linear model was proposed by Beran (1993a). Methodology for generalized linear models is well developed and implemented in standard statistical software packages (e.g., SPLUS).

In a generalized linear model we observe a random response y with mean μ and distribution function F (McCullagh and Nelder 1983). The mean μ depends on explanatory variables $u_1, ..., u_k$ through a link function ν such that

$$\nu(\mu) = \beta_o + \beta_1 u_1 + ... + \beta_k u_k. \tag{6.12}$$

Equations (6.5) and (6.6) and the central limit theorem 5.3 can also be obtained from the simplifying assumptions that

$$I(\lambda_{j,n}) \approx f(\lambda_{j,n}; \theta)\xi_j, \tag{6.13}$$

where ξ_j are independent exponential random variables with mean 1. We therefore define

$$y_{j,n} = I(\lambda_{j,n}). \tag{6.14}$$

The expected value of $y_{j,n}$ is equal to

$$\mu = f(\lambda_{j,n}; \theta). \tag{6.15}$$

Suppose now that there is a link function ν such that $\nu(\mu)$ is linear in the parameters $\theta_1, ..., \theta_M$. Thus, assume that

$$\nu(\mu) = \theta_1 u_1(\lambda) + +\theta_2 u_1(\lambda) + ... + \theta_M u_M(\lambda) \tag{6.16}$$

for suitably chosen functions $u_1, u_2, ..., u_M$. We are then in the situation of generalized linear models with y equal to the periodogram, exponential distribution function F, explanatory variables $u_1(\lambda), u_2(\lambda), ..., u_M(\lambda)$, and link function ν.

Which link function is most natural to use? Long-memory processes are characterized by the property that the spectral density is proportional to

$$\lambda^{1-2H} = e^{(1-2H)\log \lambda} \tag{6.17}$$

near the origin. A natural choice of the link function is therefore

$$\nu(\mu) = \log \mu. \tag{6.18}$$

This motivates the following class of models defined in Beran (1993a):

Definition 6.1 *Let $g : [-\pi, \pi] \to R_+$ be a positive function such that*

$$\lim_{\lambda \to 0} \frac{g(\lambda)}{\lambda} = 1$$

and $g(\lambda) = g(-\lambda)$. Define $f_o \equiv 1$, and let $f_1, f_2, ..., f_p$ be functions that are smooth in the whole interval $[-\pi, \pi]$. Also, assume that $f_k(\lambda) = f_k(-\lambda)$ and for any n, the $n^* \times (p+1)-matrix$ H with column vectors $\{f_k(2\pi/n), f_k(2\pi 2/n), f_k(2\pi 3/n), ..., h_f(2\pi n^*/n)\}^t$ $(k = 0, 1, ..., p)$ is nonsingular. Furthermore, let $\theta = (\eta_o, H, \eta_1, ..., \eta_p)$ be a real vector with $1/2 \le H < 1$. We call X_t a fractional EXP process (or an FEXP process) with short-memory components $f_1, ..., f_p$ and long-memory component g, if its spectral density is given by

$$f(\lambda; \theta) = g(\lambda)^{1-2H} \exp\{\sum_{j=0}^{p} \eta_j f_j(\lambda)\}. \tag{6.19}$$

Similarly to fractional ARIMA models, the class of *FEXP* processes is very flexible. Two classes of *FEXP* models are especially useful:

1. $g(\lambda) = |1 - e^{i\lambda}|$, $f_k(\lambda) = \cos k\lambda$ $(k = 0, 1, ..., p)$: If $H = \frac{1}{2}$, we obtain the model class proposed by Bloomfield (1973). The spectral density is a product of factors of the form $\exp(\beta_j \cos k\lambda)$. The logarithm of the spectral density is assumed to be decomposable into a finite number of cosines:

$$\log f = \eta_o + \eta_1 \cos \lambda + \eta_2 \cos 2\lambda + ... + \eta_p \cos p\lambda. \tag{6.20}$$

Every sufficiently regular function can be expressed as a Fourier series

$$\sum_{j=o}^{\infty} \eta_j \cos j\lambda. \tag{6.21}$$

Therefore, in its generality, Bloomfield's class is comparable to ARMA models: any smooth spectral density can be approximated with arbitrary accuracy. The question is mainly, how many parameters do we need to obtain a good approximation. Definition 6.1 extends these models by allowing $H = \theta_2$ to assume values above $\frac{1}{2}$. If $\theta_j = 0$ for $j \ge 3$, then we obtain a fractional ARIMA$(0, d, 0)$ model with $d = H - 1/2$.

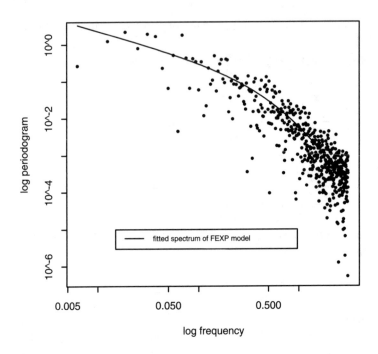

Figure 6.3. *VBR data: periodogram and fitted spectrum of polynomial FEXP model (log-log coordinates)*

2. $g(\lambda) = |1 - e^{i\lambda}|$, $h_k(\lambda) = \lambda^k$ $(k = 0, 1, ..., p)$. The logarithm of the spectral density is the sum of the long-memory component $(1 - 2H) \log \lambda$ and a polynomial in λ. A data example with $H = \frac{1}{2}$, where a second-order polynomial $(p = 2)$ makes sense intuitively, is given in Diggle (1990, p. 125 ff.). In addition, Diggle multiplied the spectrum defined by (6.19) and $H = 1/2$ by an $AR(1)$ spectrum $1/(1 - 2\alpha \cos \lambda + \alpha^2)$. The same extension can be applied here.

For illustration, we consider the VBR data. The periodogram in log-log coordinates (Figure 1.6c) has a negative slope near the origin and otherwise an essentially concave shape. A simple way to model this shape is to use the polynomial *FEXP* model (model 2 above). One would expect that, a low-degree polynomial can

Table 6.1. *Parameter estimates for the FEXP model with spectral density* $f(\lambda;\theta) = |1 - e^{i\lambda}|^{1-2H}\exp(\sum_{j=o}^{p}\eta_j|\lambda|^j)$ *fitted to the first 1000 observations of the VBR data by Heeke and Hundt (see Heeke 1991). Notation:* $\beta_o = \eta_o$, $\beta_1 = 1 - 2H$, $\beta_{j+1} = \eta_j$, $j \geq 1$. *P values are given for testing* $\beta_i = 0$ *against the two-sided alternative* $\beta_i \neq 0$.

Parameter	Estimate	Standard deviation	z-statistic	p value
$p = 2$				
β_o	−1.343	0.249	−5.39	0.0000
β_1	−0.783	0.107	−7.32	0.0000
β_2	−2.856	0.303	−9.43	0.0000
β_3	0.428	0.072	5.92	0.0000
$p = 3$				
β_o	−1.209	0.425	−2.84	0.0045
β_1	−0.741	0.150	−4.94	0.0000
β_2	−3.124	0.776	−4.03	0.0001
β_3	0.579	0.422	1.37	0.1698
β_4	−0.027	0.076	−0.36	0.7218

capture the concave shape. The results in table 6.1 and figure 6.3 * confirm this conjecture. A quadratic polynomial appears to be sufficient. There is good agreement between the observed and the fitted spectrum with $p = 2$.

## CHAPTER 7

Robust estimation of long memory

7.1 Introduction

The main assumptions in Chapters 5 and 6 were:

A1. The observed process is stationary.

A2. The observed process is Gaussian or at least a linear process.

A3. The spectral density belongs to a known parametric class of densities $f(x; \theta)$.

These assumptions are mathematically convenient. The first assumption enables us to model the process in a parsimonious way and ensures that certain fundamental mathematical properties hold (e.g., ergodicity). The mean is constant and the covariances depend on the time lag between two time points only. Thus, for n observations $X_1, ..., X_n$, there are only n unknown covariances $\gamma(k) = \text{cov}(X_t, X_{t+k})$, $k = 0, ..., n - 1$, instead of n^2 unknown covariances $\text{cov}(X_t, X_s)$, $t, s = 1, ..., n$. The assumption of Gaussianity implies that the process is fully determined by the first two moments. The likelihood function is a function of only the mean and the covariances. The same moment based estimation methods can be used for linear processes (under some moment conditions), though in general they are no longer efficient. The third assumption reduces the estimation of the spectral density function (or the covariances) to the estimation of a finite-dimensional parameter vector θ.

Mathematical convenience is useful, not only for purely theoretical considerations, but also for a straightforward intuitive understanding of the results of data analysis. Yet, in order for the above assumptions to be useful, one needs to answer the following questions:

Q1. Do assumptions A1 to A3 ever hold in practice, at least in good approximation? If not, what kind of deviations are usually

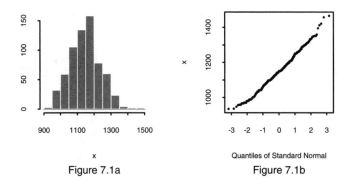

Figure 7.1a Figure 7.1b

Figure 7.1. *Histogram and normal probability plot of the Nile River minima.*

encountered?

Q2. What happens when some of the assumptions are violated? How reliable are Gaussian maximum likelihood methods in such a case?

Q3. Are there any methods that are less sensitive to deviations from the assumptions, without being too inefficient if the assumptions hold?

Let us first look at some examples.

Example 1: Consider the Nile River data. The histogram and the normal probability plot (Figure 7.1) do not reveal any relevant deviation from (univariate) normality. In the log-log plot, the periodogram follows in good approximation a straight line (Figure 6.1). Fractional Gaussian noise or a fractional ARIMA$(0,d,0)$ process therefore seems to be a good model. The maximum likelihood estimates (MLE) of H are 0.84 and 0.90, respectively. Figure 6.1 shows a good agreement between the periodogram and the fitted spectral densities. The task of modelling these data seems to be accomplished. However, taking a closer look at the time series plot of the data there appears to be a slight difference in the behavior between about the first 100 measurements and the rest of the series. The first 100 measurements seem to fluctuate around the mean in a more independent fashion. This impression is supported by a graphical comparison of the periodogram for the first 100 and

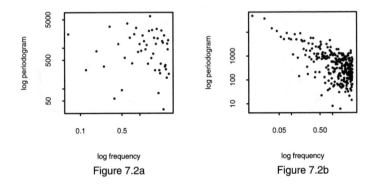

Figure 7.2a

Figure 7.2b

Figure 7.2. *Nile River minima: periodogram for* $t = 1, ..., 100$ *(Figure 7.2a) and* $t = 101, ..., 660$ *(Figure 7.2b) (in log-log coordinates).*

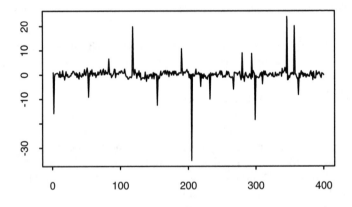

Figure   7.3.   *Simulated   series   of   a   contaminated   fractional ARIMA(0,0.3,0) process.*

the last 560 data, respectively (Figure 7.2). The question arises therefore of whether the process is nonstationary in the sense that there is a change of the dependence structure at around $t = 100$, or whether the apparent difference is likely to occur by chance for a stationary process with H around 0.9. We will see in Section 10.3 that there is indeed some evidence for a change of the long-memory parameter H.

Example 2: Suppose that the process we would like to observe is a fractional ARIMA process X_t with $H = 0.8$ and innovation variance 1. Instead of X_t, we observe

$$Y_t = (1 - Z_t)X_t + Z_t(c \cdot V_t). \tag{7.1}$$

where Z_t are independent Bernoulli variables, $P(Z_t) = 0.05$, V_t are independent t_2–distributed random variables, and $c = 10$. This means that X_t is contaminated by occasional isolated outliers. Figure 7.3 displays a typical realization of Y_t. What effect do the outliers have on MLE of H? Suppose, for instance, that the innovation variance is known to be equal to 1 and we estimate H by the approximate MLE (5.68). For 100 simulated series of length 1000, the average estimate of H turned out to be equal to 0.60, with sample standard deviation 0.044. The independent outliers V_t pull the estimate of H towards the value 0.5. Instead of estimating the dependence structure of the majority (about 95%) of the data, $\hat{H}$ reflects the independence of the occasional outliers.

Example 3: Consider the logarithm of the VBR data. Suppose that our class of models consists of fractional Gaussian noise only and we estimate H by Whittle's approximate estimator. For fractional Gaussian noise, the logarithm of the spectral density is in good approximation a linear function of the logarithm of the frequency (see Figure 2.5). Figure 6.3 shows a completely different behavior of the periodogram. Obviously, fractional Gaussian noise is not a suitable model. If we nevertheless fit fractional Gaussian noise to the data, the estimate of H is equal to 1.7 which is even outside the stationary range $(0, 1)$. By using fractional Gaussian noise, we assume a priori that the shape of the spectral density in log-log coordinates is almost a straight line. Instead, we have a curved line with very low values for high frequencies. If we fit a straight line using the periodogram ordinates at all frequencies, the estimated slope is very steep because of the low values at high frequencies. In contrast, consider the polynomial *FEXP* model with $p = 2$. Figure 6.3 shows a good agreement between observed periodogram and fitted spectral density, even for low frequencies. This good fit could be achieved by introducing onlytwo additional parameters.

The examples illustrate some typical deviations from A1 to A3. In the first case, the last two assumptions seem to hold in good approximation, whereas assumption A1 is violated. The process is not stationary, in that it appears to be divided into two different stationary parts. The second example illustrates the influence of occasional outliers. The third example shows the lack of robustness of the MLE with respect to misspecification of the model in the frequency domain. In these examples, the deviations from the original model are relatively easy to characterize. It is therefore possible to incorporate them in an improved parametric model. In other cases, this might not be possible so easily. For instance, if we observe only one outlier, the distribution of the outlier process can not be estimated. Many other kinds of deviations from A1 to A3 may occur. In the case where we firmly believe (and do not question) that observations are sampled independently from the same population distribution, the situation is much simpler. The only possible way to deviate from the model is to deviate from the marginal distribution. There is a well-developed theory of robustness for this case (see, e.g., Huber 1981, Hampel et al. 1986, Rousseeuw and Leroy 1987, Morgenthaler and Tukey 1991). The situation changes once we accept the possibility of dependence between observations at different time points. The distribution of n observations $X_1, ..., X_n$ is not determined by the marginal distribution only. In general, it is fully determined only if we know all joint distributions of $X_{t_1}, ..., X_{t_j}$ for all subsets $\{t_1, ..., t_j\} \in \{1, ..., n\}$. Therefore, there are many more possibilities of deviating from a specified model. In particular, which effect a different marginal distribution has depends on whether, and in which way, other assumptions are violated. For example, one extremely large observation can pull the sample correlation $\hat{\rho}(k)$ arbitrarily close to zero. On the other hand, if outliers occur in patches of length k, $\hat{\rho}(k)$ can be very close to 1. Also, "unusual" observations might be unusual in a more sophisticated way than just by being extremely large in absolute value. For example, a small portion of the data might consist of independent observations from a standard normal distribution, whereas the majority of the data are generated by fractional Gaussian noise with zero mean and variance 1. The outliers do not occur in the marginal distribution but rather in conditional distributions given past observations. In this case, it might be difficult to detect the unusual observations by looking at the time series plot only.

It might be unrealistic to expect that we can protect ourselves against all theoretically possible deviations from the model assumptions. Instead, one therefore considers procedures that are robust against certain types of deviations that often occur in real data. For different types of deviations different methods are proposed in the literature. None of the known methods is robust against all possible deviations. In the following sections, we give an overview of some of the known methods that focus on certain aspects of A1 to A3. A somewhat related issue is whether "long memory" in a finite data series can be generated by short-memory processes. This leads to the question of whether one should indeed use long-memory processes, instead of traditional short-memory processes for which many more theoretical results and methods are available. This is discussed in Section 7.5.

7.2 Robustness against additive outliers

A fairly general approach to characterizing the robustness properties of estimators in the time series context was given by Martin and Yohai (1986). They considered an influence functional for the so-called general replacement model. Suppose that the nominal process X_t (i.e., the process we would ideally observe) is a stationary Gaussian process for which (2.1) and (2.2) hold. As in the previous chapters, we assume that the spectral density of X_t is characterized by a finite dimensional parameter vector θ. Denote by W_t a contaminating process with distribution μ_w and by Z_t^γ a $0 - 1$ process where $0 \le \gamma \le 1$ and

$$P(Z_t^\gamma = 1) = \gamma + o(\gamma). \qquad (7.2)$$

Martin and Yohai assume that the observed contaminated series is generated by the *general replacement model*

$$Y_t^\gamma = (1 - Z_t^\gamma)X_t + Z_t^\gamma W_t. \qquad (7.3)$$

The influence of the contamination W_t on an estimator of θ can be characterized by an influence functional. Denote by μ_y^γ the distribution of the process Y_t and let $\hat\theta$ be an estimate of θ. Asymptotically, $\hat\theta$ is a functional of the distribution μ_y^γ. We therefore write $\hat\theta = \hat\theta(\mu_y^\gamma)$. Denoting the asymptotic limit of $\hat\theta(\mu_y^\gamma)$ by $\hat\theta_\infty(\mu_y^\gamma)$, the *influence functional* is defined by

$$IF(\mu_w, \hat{\theta}, \mu_y^\gamma) = \lim_{\gamma \to 0} \frac{\hat{\theta}_\infty(\mu_y^\gamma) - \hat{\theta}_\infty(\mu_y^o)}{\gamma}. \tag{7.4}$$

The influence functional measures the asymptotic influence of an infinitesimal contamination in the direction determined by Z_t^γ and W_t. If IF is bounded for a certain class of contamination distributions μ_w, then the bias of $\hat{\theta}$ is bounded within this class of contaminations, at least for sufficiently small values of γ.

An important special case of the general replacement model, is the additive outliers (AO) model with independent ("isolated") outliers,

$$Y_t = X_t + V_t, \tag{7.5}$$

where $V_t = Z_t^\gamma V_t^*$, Z_t ($t = 1, 2, ...$) are independent Bernoulli random variables with success probability γ and V_t^* is an arbitrary stochastic process with marginal distribution F_v. Note that, in the above notation, $W_t = X_t + V_t^*$. For more general outlier models, we refer the interested reader to Martin and Yohai (1986). Related approaches to robustness in time series are discussed, for example in Portnoy (1977) and Künsch (1984).

Consider now the exact and approximate Gaussian MLEs discussed in the previous chapters. They are functions of the sample correlations $\hat{\rho}(1), ..., \hat{\rho}(n-1)$. This makes them sensitive to outliers and to other deviations from the normal marginal distribution. Example 2 in Section 7.1 illustrates this lack of robustness. To obtain a bounded influence functional for AO, Beran (1993b) considered the following class of estimators. Define $\epsilon_t(\eta)$ by (5.61), $\nu_t(\theta) = \epsilon_t(\eta)/\sqrt{\theta_1}$ and

$$\dot{\nu}_t(\theta) = [\frac{\partial}{\partial \theta_1} \nu_t(\theta), ..., \frac{\partial}{\partial \theta_M} \nu_t(\theta)]^t. \tag{7.6}$$

Let $\psi = (\psi_1, ..., \psi_M)^t$ be a function from $R \times R^M \times R^M$ to R^M such that

$$\psi_i(x, y, \theta) = \psi_{i1}(x)\psi_{i2}(y) - c_i(\theta), \tag{7.7}$$

where c_i is a constant depending on θ, $x \in R, y \in R^M$, ψ_{i1} ($i = 1, ..., M$) are functions from R to R and ψ_{i2} ($i = 1, ..., M$) are functions from R^M to R. Define

$$\xi_t(\theta) = \psi(\nu_t(\theta), \dot{\nu}_t(\theta), \theta) \tag{7.8}$$

and let $\dot{\xi}_t(\theta)$ denote the derivative of $\xi_t(\theta)$ with respect to θ. Note that $\dot{\xi}_t$ is an $M \times M$-matrix. Assume that ψ is such that

$$E[\xi_t(\theta^o)] = 0 \tag{7.9}$$

and the matrices

$$A = E[\xi_t(\theta^o)\xi_t^t(\theta^o)] = E[\xi_1(\theta^o)\xi_1^t(\theta^o)] \tag{7.10}$$

and

$$B = E[\dot{\xi}_t(\theta^o)] = E[\dot{\xi}_1(\theta^o)] \tag{7.11}$$

are well defined. The estimator $\hat{\theta}$ is defined to be the solution of

$$\sum_{t=2}^{n} \psi(r_t(\theta), \dot{r}_t(\theta), \theta) = 0. \tag{7.12}$$

This is a straightforward generalization of the approximate maximum likelihood equation (5.68). For $\psi_{i1} = x, \psi_{i2}(y) = y, c_i = 0$ ($i \geq 2$) and $c_1 = -1/(2\theta_1)$, we obtain (5.68).

Whether the influence functional of $\hat{\theta}$ is bounded for the AO model depends on the function ψ. The following expression for the influence functional for the AO model follows from Martin and Yohai. Let

$$G_o = E[\psi(\epsilon_1(\theta^o) + V_1, \dot{\epsilon}_1(\theta^o), \theta^o)] \tag{7.13}$$

and

$$G_j = E[\psi(\epsilon_1(\theta^o) + b_j(\theta^o)V_1, \dot{\epsilon}_1(\theta^o) + \dot{b}_j(\theta^o)V_1, \theta^o)], \ j \geq 1. \tag{7.14}$$

Then

$$IF(\mu_w, \hat{\theta}, \mu_y^\gamma) = -B^{-1} \sum_{j=0}^{\infty} G_j. \tag{7.15}$$

In general, the number of non-zero terms on the right-hand side of (7.15) is infinite. This implies that even a bounded ψ-function does not guarantee robustness against all possible distributions μ_v. For instance, let F_v be a point mass δ_v, i.e., $V_t \equiv v$. If v tends to infinity and $\psi(x, y)$ does not converge to zero for $|x| \to \infty$ or $|y| \to \infty$, then $|IF|$ tends to infinity. To obtain a bounded influence function, a redescending ψ-function must be used. For instance, in the simplest case with known scale $\theta = H$, we may define

$$\psi_{rob}(x, y) = \psi_1(x)\psi_2(y), \tag{7.16}$$

where

$$\psi_1(x) = u(x; \alpha_1, \beta_1), \ 0 < \alpha_1 \leq \beta_1, \tag{7.17}$$

$$\psi_2(y) = u(y; \alpha_2, \beta_2), \ 0 < \alpha_2 \leq \beta_2, \tag{7.18}$$

$$u(x; \alpha, \beta) = -u(-x; \alpha, \beta) \tag{7.19}$$

and for $x \geq 0$,

$$u(x; \alpha, \beta) = x1\{x < \alpha\} + \alpha(1 - \frac{x - \alpha}{\beta - \alpha})1\{\alpha \leq x < \beta\}. \quad (7.20)$$

The resulting influence function is then bounded as v tends to infinity, provided that $\alpha_1, \alpha_2, \beta_1, \beta_2$ are finite numbers.

Recall that the R/S-statistic is also robust against deviations from the marginal distribution (theorem 4.3). The distribution properties of the R/S−statistic seem, however, rather complicated. In contrast to that, it is relatively easy to derive the following central limit theorem (Beran 1993b):

Theorem 7.1 *Suppose that X_t is a Gaussian process as defined above and $\hat{\theta}$ is defined by (7.12). Then the following holds.*

(i) There is a sequence of solutions $\hat{\theta}_n$ of (7.12) such that, as n tends to infinity,

$$\hat{\theta}_n \to \theta^o \quad (7.21)$$

almost surely.

(ii) As n tends to infinity,

$$n^{\frac{1}{2}}(\hat{\theta} - \theta^o) \to_d \zeta \quad (7.22)$$

where ζ is a normal random vector with mean zero and covariance matrix

$$V = B^{-1}AB^{-1}. \quad (7.23)$$

Note in particular that for $\psi_{i1}(x) = x$ and $\psi_{i2}(y) = y$, V is equal to the asymptotic covariance matrix of the MLE [equation (5.25)]. Using this theorem, one may find a suitable compromise between robustness against deviations from the ideal model and efficiency under this model, by defining a ψ-function for which the influence functional does not exceed a certain limit for a certain class of contaminations while the asymptotic efficiency under the ideal model is as high as possible. This is illustrated by simulations in Beran (1993b). There, the ψ−function defined by (7.16) through (7.20) is used with the tuning parameters $\alpha_1, \beta_1, \alpha_2$, and β_2 such that, for the uncontaminated model, the efficiency is equal to 0.9. If multiple roots occur, the highest value of $\hat{H}$ is accepted as solution. In spite of the high efficiency, the average estimate of H in 100 simulated series (of length $n = 1000$) of the contaminated process (7.1) with $c = 100$ turned out to be 0.799 as compared to 0.499 for the MLE.

7.3 Robustness in the spectral domain

Long memory is characterized by the behavior of the spectral density at the origin. If the slowly varying function in (2.2) converges to a constant c_f at the origin, then

$$\log f(\lambda) \approx \log c_f + (1 - 2H)\log \lambda \qquad (7.24)$$

for small frequencies λ. For simple models such as fractional Gaussian noise or a fractional $\mathrm{ARIMA}(0,d,0)$ process, this is a good approximation even for large frequencies. Fitting these models therefore basically amounts to fitting a straight line to the periodogram in log-log coordinates. Example 3 in Section 7.1 illustrates what may happen, if in contrast to our assumption, (7.24) holds in a small neighborhood of zero only. By fitting a straight line to the whole periodogram, we try to approximate a nonlinear function by a linear function. This leads to a biased estimate of H. Analogous comments apply to more complicated parametric models that do not contain the true spectral density. Misspecification of the model may lead to a serious bias in the estimate of H. There are at least three possible ways to solve this problem:

1. Use a sufficiently flexible parametric class of models. Estimate the parameters by (exact or approximate) MLE.

2. Estimate H and c_f semiparametrically by considering a decreasing proportion of smallest frequencies. As an alternative, one may consider the periodogram at all frequencies for the aggregated series

$$Y_t = \sum_{s=m(t-1)+1}^{mt} X_t,$$

where $m \geq 1$ is sufficiently large.

3. Use a simple parametric model. Estimate H and c_f by a robustified maximum likelihood method that bounds the influence of periodogram ordinates at high frequencies.

The first method leads to consistent or approximately consistent estimation of the spectral density, provided that a suitable parametric class is chosen. "Suitable" means that either the true model is contained in the model class, or the true spectral density can be approximated with high accuracy by spectral densities in the model class. For instance, every piecewise continuous function can be approximated arbitrarily well by ratios of trigonometric polynomials. Therefore, fractional $\mathrm{ARIMA}(p, d, q)$ models can approximate the spectral density of any long-memory process whose short-memory part of the spectral density is piecewise continuous. The approximation can be arbitrarily accurate, provided that large enough values of p and q are chosen. The same is true for $FEXP$ models with polynomial or trigonometric short-memory components. A question that needs to be answered for this approach is how large the dimension of the parameter vector θ should be, or more generally, which of the considered models should be used. This may be done, for instance, by applying one of the model choice criteria known in the time series literature (see, e.g., Akaike 1969, 1973a,b, 1978, 1979; Rissanen 1978, 1983; Schwarz 1978; Parzen 1974; Mallows 1973; Hannan 1980), though they are usually derived either in the context of independence or short-memory dependence. For long-memory processes, a version of Akaike's criterion based on the Whittle estimator is derived in Beran (1989b). In practice, it is not always necessary to use automatic model selection procedures. Instead, it is often sufficient to fit several candidate models that may arise naturally from the given context. A reasonable model may then be found by comparing the significance, relevance, and interpretability of the parameters. For instance, for the VBR data we observed a concave shape of the log-log spectrum. This seems to be typical for many VBR data (see Beran, Sherman, Taqqu, and Willinger 1992). A polynomial $FEXP$ model of degree 2 yields a good approximation to the observed spectrum.

In situations where only H and c_f are of interest, it seems rather wasteful to use complicated models with a high-dimensional parameter vector θ. This leads to methods 2 and 3. A version of method 2 was discussed in Section 4.6. Compared to method 1, the advantages are that only two parameters need to be estimated and no model choice is necessary. On the other hand, the problem of model choice is replaced by the problem of choosing appropriate lowest and highest frequencies, respectively, which are used for the estimation. Table 4.1 illustrates that quite different results may be obtained for different choices of the cut-off points. Also, disregarding most periodogram ordinates leads to considerable efficiency losses. The method is therefore not applicable to time series of short or moderate length. The primary objective of the method is exact asymptotic consistency. This is achieved at the cost of efficiency.

The aim of the third method is to achieve a compromise between bias (or consistency) and efficiency. At first, one considers a simple parametric model with a low-dimensional parameter vector θ. This model is used as a central model from which departures in the spectral domain are expected. To obtain estimates that will be meaningful even under deviations from the model, the parameters are estimated in a robust way. By robust we mean that deviations from the assumed spectral shape do not influence the estimates too much. The primary objective is to keep the bias of $\hat{H}$ small, but not necessarily equal to zero, while retaining reasonable precision under the ideal model. This approach was investigated by Graf (1983) for fractional Gaussian noise as the central model. There is no reason to restrict attention to this model only. We therefore generalize it to arbitrary parametric models with an M-dimensional parameter vector θ, where $M \geq 1$.

Graf considers the influence of one periodogram ordinate on $\hat{\theta}$. Suppose that $\hat{\theta}$ is a function of the periodogram ordinates

$$I_1 = I(\lambda_{1,n}), I_2 = I(\lambda_{2,n}), ..., I_{n^*} = I(\lambda_{n^*,n}) \tag{7.25}$$

with $\lambda_{j,n} = 2\pi j n^{-1}$ and n^* equal to the integer part of $\frac{1}{2}(n-1)$. We will use the notation

$$I = (I_1, ..., I_{n^*})^t. \tag{7.26}$$

We then may write

$$\hat{\theta} = \hat{\theta}(I) \tag{7.27}$$

To investigate the influence of I_k on $\hat{\theta}$, define

$$\mathcal{J}_k = \{1, 2, ..., n^*\} - \{k\} \tag{7.28}$$

to be the set of all integers between 1 and n^* excluding k. Also, denote by

$$I^{(k)} = (I_{j_1}, ..., I_{j_{n^*-1}})^t \quad (j_i \in \mathcal{J}_k) \tag{7.29}$$

the vector with components I_j excluding I_k and by

$$\hat{\theta}^{(k)} = \hat{\theta}(I^{(k)}) \tag{7.30}$$

the estimate obtained by omitting the kth periodogram ordinate I_k. The normalized influence of I_k on $\hat{\theta}$ is defined by

$$IFS(I_k, \lambda_{k,n}) = \frac{\hat{\theta} - \hat{\theta}^{(k)}}{1/n^*}. \tag{7.31}$$

We will call IFS the (empirical) *spectral influence function*. Suppose now that $\hat{\theta}$ is defined by an implicit equation

$$\sum_{j=1}^{n^*} \psi(I_j, \lambda_{j,n}, \hat{\theta}) = 0, \tag{7.32}$$

where $\psi = (\psi_1, ..., \psi_M)^t$ is a function from $R \times R \times R^M$ to R^M. It is assumed that ψ is chosen such that, under the assumed parametric model, $\hat{\theta}$ is consistent. The "one-leave-out estimate" $\hat{\theta}^{(k)}$ solves the equation

$$\sum_{j \in \mathcal{J}_k} \psi(I_j, \lambda_{j,n}, \hat{\theta}^{(k)}) = 0, \tag{7.33}$$

Under suitable regularity conditions on ψ, an approximate expression for IFS is obtained from (7.32) and (7.33) by the following Taylor expansion. By definition,

$$0 = \sum_{j \in \mathcal{J}_k} \psi(I_j, \lambda_{j,n}, \hat{\theta}) + \psi(I_k, \lambda_{k,n}, \hat{\theta}).$$

Expanding the first term around $\hat{\theta}^{(k)}$ leads to

$$0 \approx \sum_{j \in \mathcal{J}_k} \dot{\psi}(I_j, \lambda_{j,n}, \hat{\theta}^{(k)})(\hat{\theta}^{(k)} - \hat{\theta}) + \psi(I_k, \lambda_{k,n}, \hat{\theta}). \tag{7.34}$$

Here, $\dot{\psi}$ denotes the matrix

$$\dot{\psi} = [\frac{\partial}{\partial \theta_l} \psi_j]_{j,l=1,...,M}. \tag{7.35}$$

From (7.34) we get

$$n^*(\hat{\theta}^{(k)} - \hat{\theta}) = -[n^{*-1} \sum_{j \in \mathcal{J}_k}^{n^*} \dot{\psi}(I_j, \lambda_{j,n}, \hat{\theta})]^{-1} \psi(I_k, \lambda_{k,n}, \hat{\theta}). \tag{7.36}$$

For large n,

$$\frac{1}{n^*} \sum_{j \in \mathcal{J}_k}^{n^*} \dot{\psi}(I_j, \lambda_{j,n}, \hat{\theta})$$

can be replaced by

$$A = \lim_{n \to \infty} \frac{1}{n^*} \sum_{j \in \mathcal{J}_k}^{n^*} \dot{\psi}(I_j, \lambda_{j,n}, \theta^o), \tag{7.37}$$

and $\psi(I_k, \lambda_{k,n}, \hat{\theta})$ can be replaced by $\psi(I_k, \lambda_{k,n}, \theta^o)$. The spectral influence function can therefore be approximated by

$$IFS(I_k, \lambda_{k,n}) \approx A^{-1} \psi(I_k, \lambda_{k,n}, \theta^o). \tag{7.38}$$

Thus, the influence of I_k on θ is proportional to $\psi(I_k, \lambda_{k,n}, \theta^o)$.

For illustration, consider the discrete Whittle estimator (6.5), (6.6) with $m = n^*$. Equations (6.5) and (6.6) can be written in the form (7.32), with ψ defined by

$$\psi(x, y, \theta) = \dot{w}(y; \theta)\left(\frac{x}{f(y; \theta)} - 1\right), \tag{7.39}$$

where

$$w(y; \theta) = \log f(y; \theta) \tag{7.40}$$

and $\dot{w}$ is the vector of partial derivatives with respect to θ. Note that, if $\int(\log f - \log \theta_1)d\lambda$ is assumed to be equal to zero, then ψ_j ($j \geq 2$) may be simplified to

$$\psi_j(x, y; \theta) = \dot{w}_j(y; \theta)\frac{x}{f(y; \theta)}. \tag{7.41}$$

The right-hand side of (7.41) is proportional to x and thus unbounded. This means that each periodogram ordinate has an unbounded influence on $\hat{\theta}$. One extreme periodogram ordinate can distort the estimate of θ by an arbitrary amount.

To obtain an estimator that is reliable even if the true spectral density deviates from the ideal shape, a bounded ψ-function must be used. Graf (1983) suggests that one "huberizes" the standardized periodogram ordinates

$$I_j^* = \frac{I_j}{f(\lambda_{j,n}; \theta^o)}. \tag{7.42}$$

For a given number $u \in [0, 1)$, denote by q_u the $(1 - u)$-quantile of an exponential random variable with mean 1. This quantile is given by

$$q_u = -\log u. \tag{7.43}$$

Because I_j is bounded from below by zero, but not bounded from above, it is mainly important to bound the influence of large positive deviations. This can be done by replacing all standardized periodogram ordinates $I_j^* > q_u$ by q_u. This is called "huberizing," or, to indicate which quantile we are using, 'u-huberizing'. Equation (7.32) is replaced by

$$\sum_{j=1}^{n^*} \psi^{(u)}(I_j, \lambda_{j,n}, \hat{\theta}) = 0, \tag{7.44}$$

where

$$\psi^{(u)}(x, y; \theta) = \psi(x^*, y; \theta) - \mu^{(u)}(y), \tag{7.45}$$

$$x^* = \min(x, f(y; \theta)q_u) \tag{7.46}$$

and

$$\mu^{(u)}(y) = E[\psi^{(u)}(f(y; \theta)\xi, y; \theta)], \tag{7.47}$$

where ξ is a standard exponential variable. The heuristic argument for defining the huberizing constant q_u via the standard exponential distribution is that under the ideal model most of the periodogram ordinates I_j are approximately exponentially distributed with mean $f(\lambda_j; \theta)$. This is at least true in the sense that the exponential limiting distribution is obtained for each individual frequency that converges to a value in $(0, \pi)$ (see theorem 3.7). The probability of huberizing I_j^* is then approximately equal to u. In the example of the discrete Whittle estimator, we obtain the huberized ψ-function:

$$\psi^{(u)}(x, y, \theta) = \dot{w}(y; \theta)[\min(\frac{x}{f(y; \theta)}, q_u) - (1 - u)]. \qquad (7.48)$$

For $u = 0$ (no huberizing), this coincides with (7.39). Note that for a standard exponential random variable ξ,

$$E[\min(\xi, q_u)] = 1 - u. \qquad (7.49)$$

Therefore, $1-u$ instead of 1 has to be subtracted in order to achieve consistency under the ideal model. The spectral influence function is proportional to

$$\dot{w}(\lambda_{k,n}; \theta^o)[\min(\frac{I_k}{f(\lambda_{k,n}; \theta)}, q_u) - (1 - u)]. \qquad (7.50)$$

For small frequencies $\lambda_{k,n}$, the weight function $\dot{w}$ is proportional to $- \log \lambda_{k,n}$ and is thus monotonically increasing to infinity. This means that the standardized periodogram ordinates $I_k f^{-1}(\lambda_{k,n}; \theta)$ have an increasing influence as $\lambda_{k,n}$ approaches zero. This corresponds to the intuition that estimation of H should be mainly based on the behavior of the periodogram near the origin.

The main purpose of designing estimators with a bounded IFS is to keep the bias small under deviations from the model, while keeping a reasonable efficiency if the model is true. Heuristic arguments and simulations in Graf (1983) show indeed that this can be achieved. In contrast to semiparametric methods (see Section 4.6), the rate of convergence of the huberized estimator is the same as for the MLE. Generalizing Graf's result to general parametric models, the asymptotic distribution of $\hat{\theta}$ is given by

$$\sqrt{n}(\hat{\theta} - \theta^o) \to_d \kappa(u)\zeta \qquad (7.51)$$

as $n \to \infty$, where ζ is defined in Theorem 5.3 and

$$\kappa(u) = \frac{1 - u^2 + 2u \log u}{(1 - u + u \log u)^2}. \qquad (7.52)$$

This means that the standard deviation of each of the components of $\hat{\theta}$ is increased by the factor

$$\frac{\sqrt{1 - u^2 + 2u \log u}}{1 - u + u \log u}.$$

The relative efficiency of $\hat{\theta}$ as compared to the MLE can be expressed by

$$eff(u) = \kappa^{-1}(u) = \frac{(1 - u + u \log u)^2}{1 - u^2 + 2u \log u}. \tag{7.53}$$

This is a monotonically decreasing function of u. Huberizing with $u = 0$ corresponds to the approximate MLE. Therefore $eff(0) = 1$. Even for relatively large values of u, the efficiency loss is within reasonable bounds.

A simulation study in Graf (1983) illustrates the behavior of the (approximate) MLE and its huberized version. He considered, for instance, 400 realizations of length 128 of the process $Y_t = X_t + V_t$, where X_t and V_t are independent processes, X_t is fractional Gaussian noise with $H = 0.9$, and V_t is a second-order autoregressive process defined by $V_t = \alpha_1 V_{t-1} + \alpha_2 V_{t-2} + \epsilon_t$ with iid standard normal errors ϵ_t. The coefficients (α_1, α_2) considered are $(-1.83, -0.98)$ and $(-1.70, -0.85)$, respectively. In both cases, the spectral density has a local maximum near the frequency $\frac{7}{8}\pi$. In the first case, one has a sharp peak whereas in the second case the peak is rather broad. A broad peak is expected to cause more problems when one is estimating H, as more frequencies are affected. The results confirm the usefulness of the method. For instance, for model 1 and $u = 0, 0.02$ and 0.2, respectively, the average estimate of H turned out to be 0.734, 0.808, and 0.832 respectively, with sample standard deviations 0.091, 0.073, and 0.081. For model 2, the averages of the estimates were 0.725, 0.748, and 0.772, repectively, with sample standard deviations 0.074, 0.073, and 0.083. In particular, these results illustrate how the bias decreases with increasing u.

Even better results can be obtained by suitably defined frequency-dependent huberizing with a monotonically decreasing function $u(\lambda)$ and by also huberizing from below by a function $v(\lambda)$. Graf (1983) chooses, for instance,

$$u(\lambda) = \min(\frac{1}{2} + \frac{3}{8}\pi\lambda^{-1}, \frac{7}{2}) \tag{7.54}$$

and

$$v(\lambda) = 0.6\sqrt{\max(\frac{\lambda}{\pi} - 0.36, 0)}. \qquad (7.55)$$

Assuming fractional Gaussian noise as the central model, he called this estimator the *HUBINC* estimator (which stands for *HUB*erizing *INC*reases with frequency) and obtained a similiar central limit theorem as (7.51), as well as finite sample corrections for the variance of $\hat{H}$. The asymptotic variances under fractional Gaussian noise with $H = 0.5, 0.6, 0.7, 0.8$, and 0.9 are 0.519, 0.553, 0.577, 0.594, and 0.607, respectively. Other refinements considered in Graf (1983) and Graf et al. (1984) include the replacement of $f(\lambda_{j,k}; \theta)$ by the finite sample expectation of I_k and the possibility of tapering.

The refined robust analysis with the *HUBINC* estimator leads to an interesting result for the NBS data. The estimate of H turns out to be 0.602 with an approximate standard deviation of 0.044 (Graf 1983). The one-sided P-value for testing $H = \frac{1}{2}$ versus $H > \frac{1}{2}$ is about 0.01. Thus, there appears to be evidence for long memory. The approach of analyzing the dependence structure by robust estimation of H (robust in the frequency domain) is particularly suitable for this data set, since the short-memory features are difficult to interpret in a meaningful way. The reason is that the measurements were recorded at irregularly spaced time points. By treating the irregular spacing as a regular time grid, the long-term properties of the data are not destroyed. However, it is very likely that there is a considerable distortion of the short memory properties. The periodogram at higher frequencies is therefore less reliable.

As a cautionary remark, it should be noted that robustification via the spectral influence function is designed to provide robustness against arbitrary departures at a relatively small number of frequencies. The bulk of the periodogram ordinates are still expected to follow the central model. It can therefore not be expected that these methods work well if the model is grossly misspecified at the majority of the frequencies. For instance, in the case of the VBR data, the spectral density clearly has a nonlinear concave shape for most frequencies. If we use fractional Gaussian noise, whose spectral density is almost linear in log-log coordinates, none of the above methods can provide a good estimate of H.

7.4 Nonstationarity

One of the typical features of stationary long-memory processes is that there appear to be local trends and cycles, which are, however, spurious and disappear after some time. This property can make it rather difficult to distinguish a stationary process with long memory from a nonstationary process. For short time series, it might be almost impossible to decide this question. Consider, for example, the time series plots in Figures 7.4a through d. In 7.4a and 7.4c there is a clear trend downwards, whereas in 7.4b one observes a positive trend. Also, in Figure 7.4a the trend seems to be superimposed by a cyclic component. All three time series are part of the same simulated realization of fractional Gaussian noise (Figure 7.4d). The observed trends and cycles are therefore not real, but occurred "by chance." Naturally, our position here is far more comfortable than in practice - we know how the series were generated. In applications where only one of the series displayed in Figures 7.10a to 7.10c is observed, one would be led to the conclusion that the observed process is not stationary. Without any additional information, there would be no reason to believe in stationarity. Conclusions based on such short series should, however, be interpreted with caution.

Sometimes, additional, possibly non-numerical, information is available that favors either nonstationarity or stationarity. For instance, if we perform repeated measurements of the same quantity, a monotonic unbounded trend is unlikely to occur, unless the measuring equipment gets completely out of control. A typical example is the NBS data set. There, the measurements are spread over a time interval of several years. The measuring equipment was readjusted in much shorter time intervals, so that even local deterministic trends are unlikely. On the other hand, consider the temperature data given in Chapter 1. This is only one of several temperature series that indicate an increase of the temperature over about the last 100 years. Moreover, theoretical climatological models seem to support the conjecture of global warming. For the single temperature series, one might be rather uncertain of whether the apparent trend might have been generated accidentally by a stationary process (with long memory). In view of the additional scientific evidence and other temperature records, an actual trend seems to be more likely (see also the discussion in Section 9.2).

Simple persisting trends can be distinguished from stationarity, if we only wait long enough. For instance, a linear trend can eas-

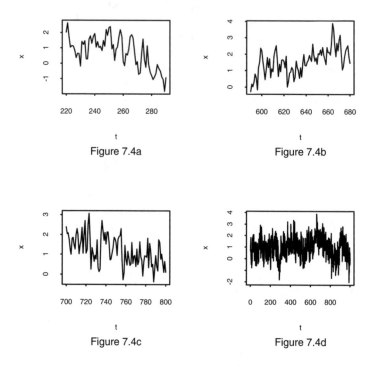

Figure 7.4. *Simulated series of a fractional ARIMA(0,0.4,0) process (Figure 7.4a) and three seemingly nonstationary parts (Figures 7.4a,b,c)*

ily be distinguished from a stationary process, if the sample size is large. The situation becomes more difficult if we allow for local trends. If we try to distinguish stationarity from arbitrary trends that persist for arbitrarily short periods, the task becomes impossible. Some restrictions need to be imposed on the nature of these trends. More generally speaking, there is a vast range of possible nonstationarities. In principle, it is therefore not possible to decide for sure whether the underlying process is indeed stationary. Ultimately, this is rather a philosophical question that might not be of primary practical interest. From the practical point of view, the assumption of stationarity provides a simple framework for parsimonious modelling. Unless there is evidence against it, it tends to be more useful than complicated nonstationary models.

More subtle types of global trends that might be confused with long memory are trends that decay to zero asymptotically. That is, we consider

$$X_t = \mu(t) + \epsilon_t, \tag{7.56}$$

where $\mu(t)$ converges slowly to zero with increasing t and X_t is a stationary process with spectral density f and $H = \frac{1}{2}$. A simple example is

$$X_t = at^{-\beta} + \epsilon_t \tag{7.57}$$

with $0 < \beta < \frac{1}{2}$. We saw in Section 4.2 that in this case the R/S-estimate of H converges to a value larger than $\frac{1}{2}$. Thus, one may be led to the erroneous conclusion that X_t is a stationary long-memory process. In contrast, exact or approximate MLE leads to the asymptotic value $\frac{1}{2}$. The reason is that the distribution of the periodogram is not influenced by the decaying trend, except in a diminishingly small neighborhood of the origin. More exactly, Künsch (1986a) shows the following result:

Theorem 7.2 *Let X_t be a Gaussian process defined by (7.56) with a bounded monotonic trend such that $\lim_{t \to \infty} \mu(t) = 0$. Then, for any $\frac{1}{2} < \gamma_1 < \gamma_2 < 1$, $\epsilon_1 > 0, \epsilon_2 > 0$ and Fourier frequencies $\lambda_{j,n}$ with*

$$\epsilon_1 n^{\gamma_1} \le j \le \epsilon_2 n^{\gamma_2}, \tag{7.58}$$

the periodogram ordinates $I(\lambda_{j,n})$ are asymptotically iid and up to a multiplicative constant χ_2^2-distributed.

Thus, apart from a few of the lowest frequencies, near the origin the periodogram should resemble the periodogram of white noise. On the other hand, for a stationary process with spectrum (2.2), Künsch showed that for small frequencies bounded from below by the left-hand side of (7.58), the periodogram behaves like a constant times a χ_2^2-variable multiplied by λ^{1-2H}. Thus, in contrast to the model (7.56), the values of the periodogram tend to increase with decreasing frequency. More exactly, the result can be summarized as follows:

Theorem 7.3 *Let X_t be a stationary Gaussian process with spectral density (2.2). Also let $j_n(1) < j_n(2) < ... < j_n(k)$ be k integer sequences such that, as $n \to \infty$,*

$$\frac{j_n(1)}{\sqrt{n}} \to \infty \tag{7.59}$$

and

$$\frac{j_n(k)}{n} \to 0. \tag{7.60}$$

Define

$$\lambda_{j_n(i)} = \frac{2\pi j_n(i)}{n}$$

and

$$I(i) = \lambda_{j_n(i)}^{-1+2H} I(\lambda_{j_n(i)}).$$

Then $I(1), ..., I(k)$ are asymptotically independent and identically distributed like a constant times a χ_2^2 random variable.

Note that λ^{1-2H} may be replaced by the spectral density itself. A difficulty in applying these results is that no general guidelines for choosing the constants $\epsilon_1, \epsilon_2, \gamma_1, \gamma_2$ are given. In some cases, however, the results can be applied qualitatively. Consider, for example, the Nile River data, The log-log plot of the periodogram shows a clear increase of the periodogram as the frequency approaches zero, over a large range of frequencies. This excludes model (7.56).

7.5 Long-range phenomena and other processes

To conclude this chapter, we discuss the general question of whether apparent "long-memory features" in a finite data set can be generated by models other than stationary processes with long memory. This complements the discussion in Section 1.5.6.

If we fix the sample size n, then the obvious answer to the above question is yes. For instance, a stationary Gaussian process with correlations $\rho(k) = \frac{1}{2}(|k+1|^{2H-2} - 2|k|^{2H-2} + |k-1|^{2H-2})$, for $k = -(n-1), -(n-2), ..., n-1$ and zero correlations for higher lags cannot be distinguished from fractional Gaussian noise. A less artificial example is an ARMA model with high correlation. Consider, for instance, an ARMA(1,0,1) process with $\phi = 0.9$ and $\psi = 0.8$. Figures 7.5a to d display the spectral density and the periodogram for the first 200, 1000, and 5000 observations of a simulated series. For small samples it is difficult to distinguish the periodogram from the periodogram of a long-memory process. Only when we observe a long series does it become apparent that the spectral density (in log-log coordinates) flattens near the origin.

As noted in the previous section, "long-memory" type behavior can also be caused by complicated global or local nonstationarities. In particular, a phenomenon that often occurs in physical sciences is that a system is initially in a nonstationary transient status and moves gradually into a stationary equilibrium. A simple example is process (7.57). It yields an alternative explanation of the Hurst effect. Fortunately, for this specific example the periodogram can

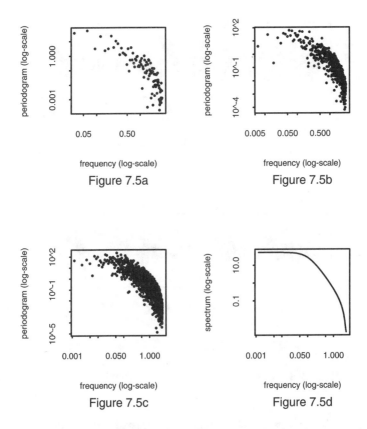

Figure 7.5a

Figure 7.5b

Figure 7.5c

Figure 7.5d

Figure 7.5. *Periodogram (in log-log-coordinates) of a simulated ARIMA(1,0,1) series with $\phi = 0.9$ and $\psi = 0.8$. The lengths of the series are $n = 200$ (Figure 7.5a), $n = 1000$ (Figure 7.5b) and $n = 5000$ (Figure 7.5c). Figure 7.5d displays the spectral density (in log-log-coordinates).*

be used to distinguish the process from a stationary process with long memory.

There are many other ways how certain "long-memory features" of a finite data set can be generated. As always in statistics, it is not possible to decide with certainty which model is correct. In fact, it is rather unlikely that any of our models is ever correct exactly. Given a finite data set, the aim can therefore hardly be to find

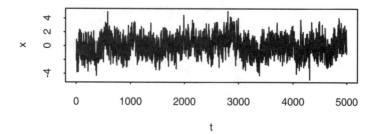

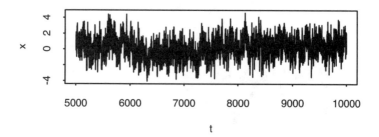

Figure 7.6. *Simulated fractional ARIMA(0,0.4,0)-series.*

the exactly correct model. Instead, one aims to find a model that
serves the intended purpose best. Criteria for choosing such models are, for example, satisfactory fit to the data, reliable prediction
of future observations, sufficiently accurate parameter estimates,
simplicity, and interpretability. In particular, an important criterion is the principle of parsimony. If the number of parameters is
relatively large compared to the number of observations, then parameter estimates and statistical inference based on the model are
very inaccurate. Moreover, complicated models with many parameters are difficult to interpret. Consider, for instance, ARMA(p, q)
models with unknown p and q. The dimensions p and q may be estimated by a suitable selection criterion that penalizes high values
of p or q. The problem of parsimony shows in the following way. If
the data generating process is a long-memory process, then with

increasing sample size, p and q tend to infinity. The reason is that we are trying to approximate the pole of the spectral density at zero by bounded continuous functions. Consider, for instance, the simulated series of a fractional ARIMA$(0,d,0)$ process with $d = 0.4$ displayed in Figure 7.6. Selecting suitable AR(p) models by estimating p via Akaike's criterion, we obtain the estimated orders $\hat{p} =$ 2, 11, 18, 21, and 25 for the first 100, 500, 2000, 4000, and 10000 observations, respectively.

In summary, it is not possible to decide with certainty whether the spectral density has a pole at zero or not, if the only information is that there is a negative slope in the log-log plot of the (finite sample) periodogram. From the applied point of view, this might be a rather philosophical question that is of secondary interest. At most one can hope that in situations where an increasing number of observations is available, one of the possibilities becomes more plausible than the other. One of the main practical arguments in favor of long-memory processes is the principle of parsimony. For finite sample sizes, the one additional parameter H allows one to model a phenomenon that would require many parameters when modeled by short-memory processes. Also, as the example of fractional ARIMA models shows, traditional short-memory processes can be embedded in a natural way as special cases in a larger class of models that allows for long memory.

## CHAPTER 8

Estimation of location and scale, forecasting

8.1 Introduction

In this chapter we consider the estimation of location and scale. We assume that the marginal distribution of the observed process X_t belongs to a location scale family of distributions defined by

$$F_{\mu,\sigma^2}(x) = F_o(\frac{x - \mu}{\sigma}). \qquad (8.1)$$

There are two main questions one may first ask:

1. What is the effect of long memory on point and interval estimates of μ and σ^2 that are used for iid data? In particular, which corrections of tests and confidence intervals are necessary in the presence of slowly decaying correlations ?

2. What are the optimal estimates?

In the following sections, questions 1 and 2 are discussed in detail. In particular, answers to (1) and (2) will tell us how much one loses by using simple standard estimates instead of more complicated optimal estimates that take the correlation structure into account.

8.2 Efficiency of the sample mean

Consider $\mu = E(X_t)$. A simple estimate of μ is given by a weighted sum of $X_1, ..., X_n$,

$$\hat{\mu}_w = \sum_{j=1}^{n} w_j X_j = w^t X, \qquad (8.2)$$

with

$$\sum_{j=1}^{n} w_j = w^t \mathbf{1} = 1. \qquad (8.3)$$

Table 8.1. *Finite sample efficiency of* $\bar{X}_n$ *for fractional Gaussian noise.*
The efficiency is defined as the ratio of $\text{var}(\bar{X}_n)$ *divided by the variance*
of the BLUE.

n	10	50	100	200
n	10	50	100	200
$H = 0.5$	1	1	1	1
$H = 0.7$	0.9898	0.9872	0.9869	0.9867
$H = 0.9$	0.9879	0.9853	0.9850	0.9848

The notation here is $X = (X_1, ..., X_n)^t$, $w = (w_1, ..., w_n)^t$, and
$\mathbf{1} = (1, 1, ..., 1)^t$. For example, the sample mean is given by (8.2)
with $w_j = n^{-1}$. Clearly, $\hat{\mu}_w$ is an unbiased estimate of μ. The
variance of $\hat{\mu}_w$ is equal to

$$\text{var}(\hat{\mu}) = \sum_{j,l=1}^{n} w_j w_l \gamma(j - l) = w^t \Sigma_n w. \tag{8.4}$$

Minimizing (8.4) under the constraint (8.3) yields

$$w = \Sigma_n^{-1} \mathbf{1} [\mathbf{1}^t \Sigma_n^{-1} \mathbf{1}]^{-1}, \tag{8.5}$$

and thus the best linear unbiased estimator (BLUE):

$$\hat{\mu} = w^t X = [\mathbf{1}^t \Sigma_n^{-1} \mathbf{1}]^{-1} \mathbf{1}^t \Sigma_n^{-1} X. \tag{8.6}$$

Its variance is equal to

$$\text{var}(\hat{\mu}) = [\mathbf{1}^t \Sigma_n^{-1} \mathbf{1}]^{-1}. \tag{8.7}$$

If X_t is a Gaussian process, then the BLUE is also the maximum
likelihood estimator of μ.

Equations (8.6) and (8.7) are simple in the sense that they pro-
vide an explicit formula for $\hat{\mu}$, and the variance of $\hat{\mu}$ respectively.
However, one needs to know all covariances $\gamma(0), \gamma(1), ..., \gamma(n - 1)$.
Usually, the covariances are unknown and have to be replaced by
estimated values. This makes (8.6) rather complicated and (8.7)
no longer holds exactly. One may therefore ask whether there are
simple alternative estimators of μ, which are almost as efficient as
the BLUE but can be calculated without knowing Σ_n. The simplest estimator is the sample mean. How much do we lose if we
use the sample mean instead of the BLUE ? The variance of $\bar{X}_n$ is
given by Theorem 2.2. The variance of the BLUE was derived by
Adenstedt (1974) (see also Samarov and Taqqu 1988, Beran and

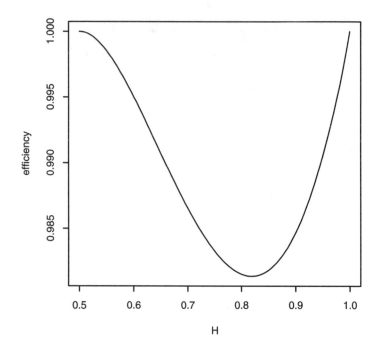

Figure 8.1. *Asymptotic efficiency of the sample mean as a function of H.*

Künsch 1985). The relative asymptotic efficiency of $\bar{X}_n$ compared to the BLUE is given by the following theorem (Adenstedt 1974).

Theorem 8.1 *The ratio of (8.7) divided by* $\mathrm{var}(\bar{X}_n)$ *is asymptotically equal to*

$$eff(\bar{X}_n, BLUE) = \frac{2H\Gamma(H + \frac{1}{2})\Gamma(3 - 2H)}{\Gamma(\frac{3}{2} - H)}. \qquad (8.8)$$

Particularly interesting is that (8.8) depends only on the parameter H. That is, only the behavior of the spectral density at the origin, and not its whole shape, determines the efficiency of the sample mean. Figure 8.1 displays $eff(\bar{X}_n, \hat{\mu}_{BLUE})$ as a function of H. For $H \geq \frac{1}{2}$, the asymptotic efficiency of $\bar{X}_n$ is always above 98%. Therefore, it is sufficient to use the sample mean instead of the much more complicated BLUE. Table 8.1 illustrates that the finite sample efficiency of $\bar{X}_n$ is high too.

Another interesting property of the sample mean was noted by Percival (1985). He considered the question of what effect leaving out every kth observation has on estimation of μ by $\bar{X}_n$. If $H > \frac{1}{2}$, then the asymptotic variance of

$$\bar{X}_n(k) = \frac{1}{m} \sum_{i=1}^{m} X_{ik}$$

with $m = [n/k]$, turns out to be independent of k. This is in sharp contrast to short-memory processes, where the variance of $\bar{X}_n(k)$ is proportional to m^{-1} and thus increases with k. The variance of $\bar{X}_n(k)$ differs aymptotically for different values of k only if a more refined comparison is made, namely via the so-called deficiency (see Hodges and Lehmann 1970). The asymptotic deficiency of $\bar{X}_n(k)$ with $k > 1$ compared to $\bar{X}_n(1)$ is zero.

8.3 Robust estimation of the location parameter

The sample mean is the optimal estimator for iid normal observations. The previous results show that, even if there is long memory in the data, the efficiency of $\bar{X}_n$ as compared to the BLUE is still very high. For normal observations, the BLUE is also the maximum likelihood estimator and is therefore optimal in the sense that, it has the smallest possible variance among all unbiased estimators. This implies that for Gaussian long-memory processes, the sample mean is almost optimal. In practice however, deviations from the normal distribution are expected. The sample mean is very sensitive to outliers and other deviations from normality. For instance, one single outlier can pull the sample mean to any arbitrary value. One therefore often considers other location estimators that are less sensitive to deviations from the ideal distribution. A large number of useful estimators can be defined as or approximated by M-estimators (see e.g. Huber 1981, Hampel et al. 1986). A definition in the context of location estimation can be given as follows. Assume that the standard deviation is known and equal to 1. This assumption is not essential but simplifies the presentation. An M-estimator $\hat{\mu}$ of μ is defined by

$$\sum_{i=1}^{n} \psi(X_i - \hat{\mu}) = 0, \tag{8.9}$$

where ψ is a function such that, for the ideal model distribution F_μ,

$$\int \psi(x - \mu)dF_\mu(x) = 0. \qquad (8.10)$$

For instance, the sample mean is defined by (8.9) with $\psi(x) = x$. The median is obtained by setting ψ equal to the sign of x. Particularly interesting are bounded ψ-functions, as the resulting estimates are not sensitive to deviations from the ideal (marginal) distribution (see, e.g., Hampel et al. 1986). In general, the price one has to pay for this security is an increased variance under the model distribution. For iid observations, $\sqrt{n}(\hat\mu - \mu)$ is asymptotically normally distributed with zero mean and variance

$$v(\psi) = \frac{E[\psi^2(X - \mu)]}{E^2[\psi'(X - \mu)]} \qquad (8.11)$$

(see, e.g., Huber, 1964, 1967, 1981, Serfling, 1980, Clarke, 1983, Hampel, 1971). Unless ψ is proportional to the score function, $v(\psi)$ is always larger than the variance of the maximum likelihood estimator. For instance, for the normal distribution, the Huber estimator has an asymptotic variance of 1.2625, 1.1073, 1.0371, and 1, when we set c equal to 0.5, 1.0, 1.5, and infinity, respectively. The value $c = \infty$ corresponds to $\bar{X}_n$.

For dependent observations with a bounded spectral density, not only the function ψ but also the exact nature of the dependence determines how much efficiency is lost by robust estimation. Gastwirth and Rubin (1975b) consider, for example, a Gaussian AR(1) process

$$X_t = \beta X_{t-1} + \epsilon_t \qquad (8.12)$$

with $-1 < \beta < 1$. They show that the efficiency of the median tends to zero as β tends to -1. In contrast, they also show that, the asymptotic efficiency of the median can be arbitrarily close to one, though below one, for positively correlated processes. Related results are also given in Zamar (1990).

In view of the latter result, one would expect that for Gaussian processes with long memory, robust M-estimators might not lose too much efficiency. It turns out that an even stronger result holds. Suppose that the observed process is defined by

$$X_t = \mu + G(Z_t), \qquad (8.13)$$

where Z_t is a stationary Gaussian process with zero mean, variance 1 and long-range dependence defined by (2.2). Furthermore, let Z

be standard normal random variable, G a real measurable function, H_k the kth Hermite polynomial,

$$a_k(\psi, G) = E[\psi(G(Z))H_k(Z)] \qquad (8.14)$$

the kth Hermite coefficient of $\psi(G)$,

$$a_k(G) = E[G(Z)H_k(Z)] \qquad (8.15)$$

the kth Hermite coefficient of G, and

$$a_o(\psi', G) = E[\psi'(G(Z))] \qquad (8.16)$$

the expected value of $\psi'(G(Z))$. We assume that $E[G(Z)] = 0$, $E[G(Z)^2] < \infty$, $a_1(G) \neq 0$, $a_1(\psi, G) \neq 0$, and $a_o(\psi', G) \neq 0$. Let

$$c_1 = \frac{a_1(\psi, G)}{a_o(\psi', G)}, \qquad (8.17)$$

$$c_2 = \frac{c_1}{a_1(G)}, \qquad (8.18)$$

and

$$\sigma_\mu = \sqrt{\frac{c_\gamma}{H(2H - 1)}}. \qquad (8.19)$$

Theorem 2.2 implies that the normalized sample mean $n^{1-H}\sigma_\mu^{-1}(\bar{X}_n - \mu)$ converges to a standard normal random variable. The limiting distribution for general M-estimators follows from Beran (1991):

Theorem 8.2 *Under mild regularity conditions on ψ, the following holds.*

(i) There exists a sequence $\hat{\mu}_n$ for which (8.9) holds and, as $n \to \infty$,

$$\hat{\mu}_n \to \mu \qquad (8.20)$$

almost surely.

(ii) As $n \to \infty$,

$$n^{1-H}\sigma_\mu^{-1}c_1^{-1}(\hat{\mu}_n - \mu) \to_d Z \qquad (8.21)$$

where Z is a standard normal random variable. Moreover,

$$n^{1-H}\sigma_\mu^{-1}[(\hat{\mu}_n - \mu) - c_2(\bar{X}_n - \mu)] \to 0 \qquad (8.22)$$

in probability.

Theorem 8.2 implies that the relative asymptotic efficiency of $\hat{\mu}$ compared to the sample mean is equal to

$$eff(\hat{\mu}, \overline{X}_n) = \lim_{n \to \infty} \frac{\text{var}(\overline{X}_n)}{\text{var}(\hat{\mu}_n)} = c_2^{-2}. \qquad (8.23)$$

In general, c_2^{-2} is not equal to 1 and both constants c_1 and c_2 depend on the function ψ. However, for the special case of Gaussian observations, i.e., $G(x) \equiv x$, the following corollary holds:

Corollary 8.1 *If $G(x) \equiv x$, then all M-estimators are asymptotically equivalent to the arithmetic mean $\overline{X}_n$, in the sense that*

$$c_2 = 1 \tag{8.24}$$

and

$$n^{-H}\sigma_\mu^{-1}(\hat{\mu}_n - \overline{X}_n) \to 0 \tag{8.25}$$

in probability.

Thus, for Gaussian long-memory processes, no efficiency is lost by robustification. The reason is that asymptotically we can use the approximation

$$\hat{\mu}_n - \mu \approx -\frac{\sum_{t=1}^{n} \psi(X_t - \mu)}{a_o(\psi', G)}. \tag{8.26}$$

From the results in Chapter 3 we know that for $H > \frac{1}{2}$ the right-hand side of (8.26) is asymptotically dominated by the first term in the Hermite polynomial expansion. For $G(x) = x$, this linear term is the sample mean. In contrast, for independent or weakly dependent processes, all terms in the expansion contribute. The asymptotic variance $v(\psi)$ is equal to

$$v(\psi) = a_o^{-2}(\psi', G) \sum_{k=1}^{\infty} \frac{a_k^2(\psi, G)}{(k!)^2} = \frac{E[\psi^2(Z)]}{E^2[\psi'(Z)]}. \tag{8.27}$$

The asymptotic efficiency is then equal to

$$v^{-1}(\psi) = \frac{E^2[\psi'(Z)]}{E[\psi^2(Z)]}. \tag{8.28}$$

This expression is smaller than 1, unless $\psi(x)$ is proportional to x. For instance, for the Huber estimator defined above, $v^{-1}(\psi)$ is monotonically decreasing in c, starting at the value of 1 for $c = \infty$, and approaching $2/\pi$ for $c \to 0$.

Beran (1991) also considered the special case of short-memory processes where the sum of all correlations is zero. Although under long-range dependence we do not lose any efficiency by using a robust estimator, exactly the opposite happens here. The asymptotic efficiency of all M-estimators with a nonlinear ψ-function is zero. The reason is that the variance of the sample mean converges to zero at a faster rate than n^{-1}. This follows from the fact that

Table 8.2. *Variance of the median and the Huber-estimator with c = 1.5.*
The results are based on 20000 simulations of fractional Gaussian noise
with mean zero and variance one.

	n=16 Median	Huber	n=64 Median	Huber	n=256 Median	Huber
$H = 0.5$	1.452	1.035	1.540	1.038	1.556	1.043
$H = 0.7$	1.141	1.011	1.088	0.978	1.046	1.006
$H = 0.9$	1.040	1.006	1.013	0.996	1.010	0.998

var$(\bar{X}_n)$ is equal to the sum of covariances $\gamma(i - j)$ $(i, j = 1, ..., n)$ and the assumption that the sum of all covariances is zero. This link is destroyed when a nonlinear ψ-function is used. The variance of the resulting estimator converges to zero at the rate n^{-1}.

To illustrate the asymptotic results, Table 8.2 * gives simulated normalized variances for the median and the Huber estimator with $c = 1.5$. The simulated process is fractional Gaussian noise with $\mu = 0$ and $\sigma = 1$. For $H = 0.5$, the asymptotic variance is equal to 1 for $\bar{X}$, 1.037 for the Huber estimator, and $\pi/2 \approx 1.571$ for the median. For $H > \frac{1}{2}$, the asymptotic variance of all three estimators is equal to 1. Note that for the sample mean, the normalized finite sample variance is also exactly equal to 1. The results indicate that, if the long-range dependence is very strong, the normalized variance stabilizes faster around the asymptotic value of 1, than for less strong dependence. The theoretical explanation is that the linear term in the expansion of $\hat{\mu}_n - \mu$ is of the order n^{H-1}. For high values of H, this term dominates the nonlinear terms faster. On the other hand, for H close to $\frac{1}{2}$, the nonlinear term becomes more important for finite samples. For $1/2 \leq H < 3/4$, it can be shown by similar arguments as above that the nonlinear term is of the order $n^{-\frac{1}{2}}$. For $H = \frac{1}{2}$, this is also the order of the linear term.

* Adapted from Table 1 of Jan Beran (1991) "M-estimators of location for Gaussian and related processes with slowly decaying serial correlations." JASA, Vol. 86, No. 415, 704-708, with the permission of the author and the publisher.

8.4 Estimation of the scale parameter

The most common estimator of variance σ^2 of a random variable X is the sample variance

$$s^2 = \frac{1}{n-1} \sum_{t=1}^{n} (X_t - \bar{X})^2. \tag{8.29}$$

In Section 1.1, we obtained the general expression

$$E(s^2) = \sigma^2 [1 - (n-1)^{-1} \delta_n(\rho)]. \tag{8.30}$$

Table 1.4 illustrates that for large values of H, the bias term $\Delta_n = -(n-1)^{-1} \delta_n(\rho)$ converges to zero rather slowly, as n increases. If we are able to estimate the correlation structure, then an (almost) unbiased estimator of σ^2 can be obtained by multiplying s^2 with the corresponding estimated correction factor $1 - \hat{\Delta}_n$. That is, we may define the new scale estimator

$$\tilde{s}^2 = (1 - \hat{\Delta}_n) s^2. \tag{8.31}$$

For instance, for fractional Gaussian noise, this leads to

$$\tilde{s}^2 = \frac{n - n^{2\hat{H}-1}}{n-1} s^2, \tag{8.32}$$

where $\hat{H}$ is a consistent estimate of H. This estimator is not exactly unbiased hpwever, because $\hat{H}$ is estimated.

One advantage of s^2 for iid normal observations is that the distributional properties of the sample variance are very simple. Multiplied by $n-1$, s^2 is exactly distributed like σ^2 times a χ^2-variable with $n-1$ degrees of freedom. This simplicity does not carry over to the case of dependent observations. For long-memory processes with $H > 3/4$, even the rate at which s^2 converges to σ^2 is different. This can be seen by writing

$$(n-1)(s^2 - \sigma^2) = \sigma^2 \sum_{t=1}^{n} \left[\frac{(X_t - \mu)^2}{\sigma^2} - 1 \right] - n(\bar{X} - \mu)^2.$$

The first term is a sum of the second Hermite polynomial applied to $(X_1 - \mu)/\sigma$, ..., $(X_n - \mu)/\sigma$ respectively. The second term is of the order $O_p(n^{2H-1})$. For $\frac{1}{2} < H < \frac{3}{4}$, the first term dominates and is asymptotically equal to $\sqrt{n}$ times a normal random variable (see Theorem 3.1). For $\frac{3}{4} < H < 1$, n^{1-2H} times the first term converges to a non-normal random variable. The second term multiplied by a constant times n^{1-2H} converges to a χ^2-variable with one degree of freedom. Thus, although the bias of the sample variance can be taken into account by (8.31), even the corrected estimator remains rather inaccurate and its distributional properties are complicated. This is in contrast to estimators obtained by maximum likelihood and related methods discussed in the previous chapters. They are $\sqrt{n}$ consistent for all $H \in [\frac{1}{2}, 1)$ and asymptotically normal.

8.5 Prediction of a future sample mean

In some situations, it is of interest to predict the average value of the next m observations. The statistical problem can be formulated as follows. Given the past values $X_{1-n}, ..., X_o$, predict the future mean

$$\bar{X}_{+m} = \frac{1}{m} \sum_{t=1}^{m} X_t. \tag{8.33}$$

For the special case where the number of past observations is infinite, the asymptotic prediction error of the best unbiased linear predictor $\hat{X}_{+m}$ is given by (Beran 1989a):

Theorem 8.3 *Under assumption* (2.2),

$$\lim_{n \to \infty} E[(\bar{X}_{+m} - \hat{X}_{+m})^2 | X_s, s \leq 0] = \frac{c_f}{2H\Gamma^2(H + \frac{1}{2})}. \tag{8.34}$$

The more general situation where the prediction is based on a finite number of observations is more diffcult to deal with. There are different kinds of limiting behaviors one can consider. For instance, m and n can be assumed to tend to infinity such that the ratio mn^{-1} tends to a positive constant in the interval $(0, 1)$. Other possibilities include $mn^{-1} \to 0$ and $mn^{-1} \to 1$. The resulting expressions may become rather messy. Also, a practical problem with the BLUE is that it depends on the unknown correlation structure. Instead of deriving theoretical results for the BLUE, we may therefore consider the simpler, though not necessarily exactly optimal predictor

$$\hat{X}_{+m} = E[\bar{X}_{+m} | \bar{X}_n]. \tag{8.35}$$

Thus, instead of conditioning on the individual observations X_1, ..., X_n we are conditioning on the observed sample mean $\bar{X}_n$. Denote by σ_n^2 and σ_m^2 the variances of $\bar{X}_n$ and $\bar{X}_{+m}$, respectively, and by $\rho_{n,m}$ the correlation between these two random variables. The mean square prediction error $E[(\bar{X}_{+m} - \hat{X}_{+m})^2 | \bar{X}_n]$ is minimized by the conditional expectation

$$E[\bar{X}_{+m} | \bar{X}_n] = \mu + \rho_{n,m} \frac{\sigma_m}{\sigma_n} (\bar{X}_n - \mu). \tag{8.36}$$

The mean square prediction error is then equal to the conditional variance

$$E[(\bar{X}_{+m} - E[\bar{X}_{+m}|\bar{X}_n])^2|\bar{X}_n] = \sigma_m^2(1 - \rho_{n,m}^2). \tag{8.37}$$

Note, however, that in (8.36) the expected value μ needs to be known. Replacing μ in (8.36) by $\bar{X}_n$ leads to the simple prediction

$$\hat{X}_{+m} = \bar{X}_n. \tag{8.38}$$

The prediction error is then equal to

$$E[(\bar{X}_{+m} - \bar{X}_n)^2|\bar{X}_n] = \sigma_m^2(1 - \rho_{n,m}^2) + (1 - \rho_{m,n}\frac{\sigma_m}{\sigma_n})^2(\bar{X}_n - \mu)^2. \tag{8.39}$$

Because μ is unknown, the second term cannot be calculated explicitly. We therefore replace $(\bar{X}_n - \mu)^2$ by the unconditional expected value

$$E[(\bar{X}_n - \mu)^2] = \sigma_n^2. \tag{8.40}$$

The uncertainty of the prediction $\hat{X}_{+m} = \hat{X}_n$ is therefore assessed by the unconditional variance

$$E[(\bar{X}_{+m} - \bar{X}_n)^2] = \sigma_n^2 + \sigma_m^2 - 2\rho_{n,m}\sigma_m\sigma_n. \tag{8.41}$$

If the observed $\bar{X}_n$ is close to μ, then this is larger than the conditional prediction error (8.39). On the other hand, if $\bar{X}_n$ is far from μ, then (8.41) is smaller than (8.39).

Expression (8.41) can be simplified by considering its asymptotic limit. Three situations can be distinguished:

1. $n \to \infty$ and $mn^{-1} \to 0$: In practice, this is applicable if n is very large compared to m. Because, σ_n is converging to zero monotonically, we obtain

$$E[(\bar{X}_{+m} - \bar{X}_n)^2] \approx \sigma_m^2. \tag{8.42}$$

If m also tends to infinity, then

$$E[(\bar{X}_{+m} - \bar{X}_n)^2] \approx \sigma_\mu^2 m^{2H-2}. \tag{8.43}$$

Using this kind of asymptotics means that $\bar{X}_n$ can be considered to be almost equal to μ. Prediction of the future sample mean $\bar{X}_{+m}$ by $\bar{X}_n$ essentially amounts to prediction by μ.

2. $m \to \infty$ and $mn^{-1} \to \infty$: This is the opposite case with n small compared to m. Asymptotically, $\bar{X}_{+m}$ can be considered to be equal to μ, so that prediciton of the future mean reduces to estimation of μ. By analogous arguments as above,

$$E[(\bar{X}_{+m} - \bar{X}_n)^2] \approx \sigma_\mu^2 n^{2H-2}. \tag{8.44}$$

3. $n \to \infty$, $m \to \infty$ and $mn^{-1} \to q \in (0,1)$: In this case, all terms in (8.41) are of the same order of magnitude. The asymptotic limit of (8.41) is given by the following theorem (Beran 1989a).

Theorem 8.4 *Let*

$$q = \lim_{n,m\to\infty} \frac{m}{n} \in (0,1).$$

Then

$$\lim_{n,m\to\infty} n^{2-2H}(\sigma_n^2 + \sigma_m^2 - 2\rho_{m,n}\sigma_n\sigma_m)$$

$$= \sigma_\mu^2\{1 + q^{2H-2} - \frac{1}{q}[(q+1)^{2H} - (q^{2H}+1)]\}. \qquad (8.45)$$

Consider, for instance, fractional Gaussian noise with mean μ and variance σ^2. The variance of $\bar{X}_n$ and $\bar{X}_{+m}$, respectively, is exactly equal to $\sigma^2 n^{2H-2}$ and $\sigma^2 m^{2H-2}$, respectively. Thus, in case 1:

$$E[(\bar{X}_{+m} - \bar{X}_n)^2] \approx \sigma^2 m^{2H-2},$$

and in case 2,

$$E[(\bar{X}_{+m} - \bar{X}_n)^2] \approx \sigma^2 n^{2H-2}.$$

For case 3, the asymptotic equation (8.45) even holds exactly for all $m = qn$ where q or q^{-1} is an integer. This is due to the self-similarity of fractional Brownian motion.

8.6 Confidence intervals for μ and a future mean

8.6.1 Tests and confidence intervals for μ with known long-memory and scale parameters

In Section 8.2 we saw that slowly decaying correlations do not have any major effect on the efficiency of the sample mean. The slow rate of convergence of $\bar{X}_n$ does, however, have a strong impact on tests and confidence intervals for μ. The z-statistic

$$z = \frac{\bar{X}_n - \mu}{\sigma n^{-\frac{1}{2}}} \qquad (8.46)$$

diverges to infinity, in the sense that

$$\lim_{n\to\infty} P(|z| > c) = 1 \qquad (8.47)$$

for any constant c. The same is true for Student's t-statistic

$$t = \frac{\bar{X}_n - \mu}{s n^{-\frac{1}{2}}}, \qquad (8.48)$$

with s^2 denoting the sample variance. The asymptotic coverage probability of confidence intervals of the form

$$\bar{X} \pm z_{\frac{\alpha}{2}} \frac{s}{\sqrt{n}} \tag{8.49}$$

is therefore equal to zero, instead of $(1 - \alpha)$. Table 1.3 in Section 1.1 illustrates that this effect of long memory is seen already for moderately large sample sizes. For tests of the null hypothesis $H_o : \mu = \mu_o$, this implies that the null hypothesis is rejected with asymptotic probability 1, even if H_o is true.

In view of Theorem 2.2, the statistic

$$z(\mu, c_f, H) = \frac{\bar{X}_n - \mu}{\sigma_\mu} n^{1-H}, \tag{8.50}$$

with σ_μ defined by (8.19), should be used instead of (8.46). Asymptotically, $z(\mu, c_f, H)$ is standard normal. Confidence intervals for μ with asymptotic coverage probability $1 - \alpha$ are given by

$$\bar{X} \pm \sigma_\mu n^{H-1}. \tag{8.51}$$

For fractional Gaussian noise, $z(\mu, c_f, H)$ reduces to

$$\frac{\bar{X}_n - \mu}{\sigma} n^{1-H} = \frac{\bar{X}_n - \mu}{\sigma n^{-\frac{1}{2}}} n^{\frac{1}{2}-H},$$

where $\sigma^2 = \mathrm{var}(X_t)$.

8.6.2 Tests and confidence intervals for a future mean, with known long-memory and scale parameters

Comments analogous to those above, apply to prediction intervals of a future sample mean $\bar{X}_{+m}$ based on the observed sample mean $\bar{X}_n$. If the observations are independent, then the variance of $\bar{X}_{+m} - \bar{X}_n$ is equal to the sum of the individual variances:

$$\mathrm{var}(\bar{X}_{+m} - \bar{X}_n) = \mathrm{var}(\bar{X}_{+m}) + \mathrm{var}(\bar{X}_n) = \sigma^2 (\frac{1}{m} + \frac{1}{n}). \tag{8.52}$$

A prediction interval for $\bar{X}_{+m}$ is given by

$$\bar{X}_n \pm z_{\frac{\alpha}{2}} \sigma \sqrt{\frac{1}{n} + \frac{1}{m}}. \tag{8.53}$$

This also defines the acceptance region for testing for equality of $\mu_1 = E(\bar{X}_n)$ and $\mu_2 = E(\bar{X}_{+m})$ against the alternative $\mu_1 \neq \mu_2$. Under long-range dependence, the asymptotic coverage probability of (8.53) is zero. The asymptotic level of significance is 1. If m and n

are of comparable orders of magnitude, then Theorem 8.4 suggests using the statistic

$$z_{n,m}(c_f, H) = \frac{\bar{X}_n - \bar{X}_{+m}}{\sigma_{\mu_1,\mu_2}(c_f, H; q)} n^{1-H} \qquad (8.54)$$

where

$$\sigma_{\mu_1,\mu_2}(c_f, H; q) = \sigma_\mu^2 \{1 + q^{2H-2} - \frac{1}{q}[(q+1)^{2H} - (q^{2H} + 1)]\}, \qquad (8.55)$$

and $q = mn^{-1}$. Asymptotically, $z_{n,m}(c_f, H)$ is standard normal. An approximate $(1 - \alpha)$-prediction interval is therefore given by

$$\bar{X}_n \pm z_{\frac{\alpha}{2}} \sigma_{\mu_1,\mu_2}(c_f, H; q) n^{H-1}. \qquad (8.56)$$

8.6.3 Tests and confidence intervals for μ with unknown long-memory and scale parameters

In practice, H and c_f usually have to be estimated. Therefore, instead of $z(\mu, c_f, H)$ one needs to consider $z(\mu, \hat{c}_f, \hat{H})$ where $\hat{c}_f$ and $\hat{H}$ are suitable estimates of c_f and H (Beran 1986, 1989a). Suppose that $\hat{H}$ and $\hat{c}_f$ converge to the correct values H and c_f, respectively. Then, for very large sample sizes, $z(\mu, \hat{c}_f, \hat{H})$ is approximately standard normal and one can use the results in Section 8.6.1. For moderate and small sample sizes, the additional variability introduced by estimating H and c_f needs to be taken into account. In particular, because H occurs in the exponent of n, even relatively small changes of H can make a big difference. A simple approximation to the distribution of z can be obtained by Taylor expansion. At first we introduce the notation $\nu = (c_f, H)$ and define

$$c(\nu; n) = \sigma_\mu^2(c_f, H) n^{2H-2}. \qquad (8.57)$$

Then

$$z(\mu, c_f, H) = z(\mu, \nu) = \frac{\bar{X}_n - \mu}{\sqrt{c(\nu, n)}}$$

and

$$z(\mu, \hat{\nu}) \approx z(\mu, \nu) + \frac{\partial}{\partial \nu_1} z(\mu, \nu)(\hat{\nu}_1 - \nu_1) + \frac{\partial}{\partial \nu_2} z(\mu, \nu)(\hat{\nu}_2 - \nu_2). \qquad (8.58)$$

The derivative of n^{1-H} is equal to $-\log n \cdot n^{1-H}$ and is thus of larger order than n^{1-H}. The two leading terms in (8.58) are therefore

$$z^{(1)} = z(\mu, c_f, H)[1 - \log n \cdot (\hat{H} - H)]. \qquad (8.59)$$

Table 8.3. *Quantiles of $Z_n^{(1)}$, obtained from 10000 simulations of Z_1 and Z_2 (except for $n = \infty$). The quantiles q are such that $P(|Z_n^{(1)}| > q) = \alpha$.*

n	$\alpha = 0.1$	$\alpha = 0.05$	$\alpha = 0.01$
100	1.745	2.237	3.145
200	1.707	2.154	2.967
400	1.685	2.080	2.878
1000	1.659	2.013	2.732
5000	1.643	1.950	2.617
∞	1.645	1.960	2.576

Suppose that $\hat{\nu}$ is one of the $\sqrt{n}$-consistent estimators discussed in the earlier chapters. In first approximation, $\bar{X}_n$ can be considered to be independent of $\hat{H}$, as $\hat{H}$ is essentially a function of the second Hermite polynomials of X_t whereas $\bar{X}_n$ is n^{-1} times the sum of the first Hermite polynomial applied to X_t (Beran 1989a). The statistic $z^{(1)}$ is therefore approximately distributed like

$$Z_n^{(1)} = Z_1(1 + \frac{\log n}{\sqrt{n}}\sigma_H Z_2), \qquad (8.60)$$

where Z_1, Z_2 are independent standard normal variables and σ_H^2 is the asymptotic variance of $\hat{H}$. This approximation is particularly simple if σ_H^2 does not depend on any unknown parameters. For instance, for a fractional ARIMA$(0, d, 0)$ model, the asymptotic variance of the (approximate) MLE of $H = d - \frac{1}{2}$ is equal to $6\pi^{-2}$. Thus, $z(\mu, \hat{c}_f, \hat{H})$ is approximately distributed like

$$Z_n^{(1)} = Z_1(1 + \frac{\log n}{\sqrt{n}}\frac{\sqrt{6}}{\pi}Z_2). \qquad (8.61)$$

Table 8.3 gives some quantiles of $Z_n^{(1)}$ for several values of n. As n increases, the quantiles approach those of the standard normal distribution. How well the desired coverage probabilities and levels of significance respectively are approximated is illustrated in table 8.4. For 400 simulated series with $H = 0.8$ and $n = 200$, empirical probabilities of $|z(\mu, \hat{c}_f, \hat{H})|$ exceeding the standard normal quantiles and the quantiles based on (8.61), respectively, are given. The levels of significance are 0.1, 0.05, and 0.01. For comparison, also the corresponding probabilities for $|z(\mu, c_f, H)|$ using standard normal quantiles are given. The quantiles for the case where c_f and H are known are obviously very well approximated

Table 8.4. *Simulated rejection probabilities for testing* H_o : $\mu = \mu_o$ *against* H_o : $\mu \neq \mu_o$. *The rejection regions are based on* $z(\mu, \hat{c}_f, \hat{H})$ *with standard normal quantiles (method 1) and quantiles based on* (8.61) *(method 2), respectively. Also given are the corresponding rejection probabilities for the case where* c_f *and* H *are known and standard normal quantiles are used (method 3). The results are based on 400 series of a fractional ARIMA(0,0.3,0) process with* $n = 200$.

method	$\alpha = 0.1$	$\alpha = 0.05$	$\alpha = 0.01$
1	0.168	0.120	0.053
2	0.155	0.095	0.025
3	0.103	0.053	0.008

by those of the standard normal distribution. This is not the case if the parameters are estimated. Approximation (8.61) yields better quantiles though they still appear to be too small. Depending on how exactly one needs to approximate the coverage or rejection probability, one may want to consider better approximations. For instance, instead of expanding $c(\hat{\nu}, n)$, Beran (1989a) suggested using the direct approximation

$$c(\hat{\nu}, n) \approx c(c_f + \frac{1}{\sqrt{n}}\sigma_{c_f} Z_3, H + \frac{1}{\sqrt{n}}\sigma_H Z_2; n), \qquad (8.62)$$

where Z_3, Z_2 are standard normal, $\sigma_{c_f}^2$ and σ_H^2 are the asymptotic variances of $\hat{c}_f$ and $\hat{H}$, respectively, and the vector $(\sigma_{c_f} Z_3, \sigma_H Z_2)$ has the joint normal distribution given by the central limit theorem for $(\hat{c}_f, \hat{H})$. This approximation is somewhat more tedious. A simplification that is more precise than (8.60) but less complicated than (8.62) can be obtained by taking into account the variability of $\hat{H}$ in the factor $n^{1-\hat{H}}$ only, without simplifying the function n^x any further. This means that we approximate the distribution of $z(\mu, \hat{c}_f, \hat{H})$ by the distribution of

$$Z_n^{(2)} = Z_1 \cdot n^{Z_2 \sigma_H n^{-\frac{1}{2}}} \qquad (8.63)$$

with independent standard normal random variables Z_1, Z_2. Note that (8.60) is obtained from (8.63) by considering the first two terms in the Taylor expansion of the function n^x only. Table 8.5 gives the resulting empirical rejection probabilities for the same simulated series as in Table 8.4. They are close to the desired probabilities. Quantiles for several sample sizes are given in Ta-

Table 8.5. *Simulated rejection probabilities as defined in table 8.3 but based on approximation (8.63).*

$\alpha = 0.1$	$\alpha = 0.05$	$\alpha = 0.01$
0.138	0.063	0.005

Table 8.6. *Quantiles of the random variable $Z_n^{(2)}$ obtained from 10000 simulations of Z_1 and Z_2 (except for $n = \infty$). The quantiles q are such that $P(|Z_n^{(2)}| > q) = \alpha$.*

n	$\alpha = 0.1$	$\alpha = 0.05$	$\alpha = 0.01$
100	1.963	2.592	4.250
200	1.842	2.397	3.626
400	1.766	2.243	3.255
1000	1.707	2.126	2.925
5000	1.661	1.993	2.676
∞	1.645	1.960	2.576

ble 8.6. Beran (1989a) also considered frequency-robust estimation of c_f and H (see Chapter 7). This is particularly appropriate in the given context, since c_f and H are the only parameters of interest. They need to be estimated in a reliable way, even if the assumed spectral shape is not quite correct at high frequencies. As another alternative, one might use fully consistent semiparametric estimation as discussed in Sections 4.6 and 7.3. Due to the large variability of such estimators, this is suitable for long series only.

8.6.4 Tests and confidence intervals for a future mean, with unknown long-memory and scale parameters

Prediction intervals for a future sample mean, with estimated c_f and H, can be obtained in a manner analogous to that described above. The details are therefore left as an exercise.

8.7 Forecasting

In contrast to estimation of constants, forecasting becomes easier the more future observations depend on the past (cf. remarks in Section 1.2). For long-memory processes, good short- and long-term predictions can be obtained when a long record of past values is available. This is illustrated below (see also the example in section 1.2). At first, we recall some standard results on prediction for linear processes.

Suppose that X_t is a linear process (3.12) with long memory, mean zero, and variance σ^2. We assume that X_t can also be represented as an infinite autoregressive process (5.6). Thus, X_t has the representations

$$X_t = \sum_{s=1}^{\infty} b(s) X_{t-s} + \epsilon_t$$

and

$$X_t = \sum_{s=0}^{\infty} a(s) \epsilon_t$$

with zero mean identically distributed uncorrelated innovations ϵ_t. We will use the notation

$$X^{(n)} = (X_1, ..., X_n)^t \qquad (8.64)$$

for the vector of observed values and

$$\gamma_k^{(n)} = [\gamma(n+k-1), \gamma(n+k-2), ..., \gamma(k)]^t \qquad (8.65)$$

for the vector of covariances between the components of $X^{(n)}$ and a future observation X_{n+k}. The simplest way to predict X_{n+k} is to take a suitable linear combination of past values

$$\hat{X}_{n+k}(\beta) = \beta^t X^{(n)} \qquad (8.66)$$

where

$$\beta = (\beta_1, ..., \beta_n)^t. \qquad (8.67)$$

The mean squared prediction error is then equal to

$$
\begin{aligned}
MSE_k(\beta) &= E[(X_{n+k} - \hat{X}_{n+k})^2] \\
&= \sigma^2 + \beta^t \Sigma_n \beta - 2\beta^t \gamma_k^{(n)}.
\end{aligned}
\qquad (8.68)
$$

Here, Σ_n denotes the covariance matrix of $X^{(n)}$. $MSE_k(\beta)$ is minimized by

$$\beta_{opt} = \Sigma_n^{-1} \gamma_k^{(n)}. \qquad (8.69)$$

The best linear prediction is therefore given by

$$\hat{X}_{n+k}(\beta_{opt}) = [\gamma_k^{(n)}]^t \Sigma_n^{-1} X^{(n)}. \qquad (8.70)$$

For Gaussian processes, this is equal to the conditional expectation of X_{n+k} given $X^{(n)}$ and is therefore best among all possible (linear

Table 8.7. $R_k^2(\beta_{opt})$ for a fractional $ARIMA(0, d, 0)$ process and the best linear prediction of X_{n+k} based on the observations $X_n, ..., X_1$.

k	$n = 1$ $H = 0.6$	$H = 0.9$	$n = 10$ $H = 0.6$	$H = 0.9$	$n = 100$ $H = 0.6$	$H = 0.9$
1	0.0123	0.4444	0.0182	0.5092	0.0190	0.5162
10	0.0003	0.1794	0.0013	0.2710	0.0021	0.3059
20	0.0001	0.1360	0.0005	0.2164	0.0011	0.2609
100	$8 \cdot 10^{-6}$	0.0714	$5 \cdot 10^{-5}$	0.1202	0.0002	0.1702

and nonlinear) predictions. The corresponding mean squared error is equal to

$$MSE_k(\beta_{opt}) = \sigma^2 - [\gamma_k^{(n)}]^t \Sigma_n^{-1} \gamma_k^{(n)}. \qquad (8.71)$$

Recursive algorithms are available for the calculation of β_{opt} and $MSE_k(\beta_{opt})$. This is discussed in detail, for instance, in Brockwell and Davis (1987). A standardized measure of how well we predict X_{n+k} may be defined as

$$R_k^2(\beta) = \frac{\sigma^2 - MSE_k(\beta)}{\sigma^2} = 1 - \frac{MSE_k(\beta)}{\sigma^2}. \qquad (8.72)$$

(see e.g. Granger and Joyeux 1980). For β_{opt}, this is the proportion of the variance of X_t that is explained by the best prediction. For a perfect prediction, MSE_k is zero and R_k^2 is equal to 1. On the other hand, if X_{n+k} does not depend on $X^{(n)}$, then the best prediction is the unconditional mean 0. Thus, no improvement can be achieved by using the observed values for prediction. Therefore, $MSE_k = \sigma^2$ and $R_k^2 = 0$. Also note that as k tends to infinity, R_k^2 tends to zero. This reflects the fact that forecasts of the remote future become more and more difficult.

Numerical values of $R_k^2(\beta_{opt})$ are given in Table 8.7 for the fractional $ARIMA(0, d, 0)$ process with $d = H - \frac{1}{2} = 0.1$ and 0.4 respectively, $k = 1, 10, 20, 100$, and $n = 1, 10$, and 100. The results illustrate that the precision of forecasts can be improved considerably by using all observed values. This is in particular true for long-term forecasts and when H is large. There the property of long memory can be used most effectively to improve the precision of forecasts. For instance, for $H = 0.6$, the value of R_k^2 for the best 20-steps-ahead forecast based on the last 100 observations is larger by a factor of about 10 and 2 than the corresponding values for the prediction based on the last and the last 10 observations, respectively. Also note that for $H = 0.9$, R_{20}^2 is about 237 times larger than for $H = 0.6$.

In view of these results, it is also interesting to compare predictions of X_{n+k} based on an infinite past $X_s(s \leq n)$ with predictions that are based on a small fixed number of observations. For simplicity, we focus on the comparison with the prediction based on the last observation X_n only. Essentially the same comments carry over to comparisons with predictions based on an arbitrary fixed number of past observations. In extension of the above notation we write

$$X^{(\infty)} = (X_s, s \leq n)^t = (X_n, X_{n-1}, X_{n-2}, \ldots)^t \qquad (8.73)$$

for the infinite dimensional vector of observations $X_s(s \leq n)$. Among all linear predictions of X_{n+1},

$$\hat{X}_{n+1}(\beta^{(\infty)}) = \sum_{s=1}^{\infty} \beta_s^{(\infty)} X_{n+1-s} = [\beta^{(\infty)}]^t X^{(\infty)}, \qquad (8.74)$$

the mean squared error is minimized by

$$\hat{X}_{n+1} = [\beta_{opt}^{(\infty)}]^t X^{(\infty)}, \qquad (8.75)$$

with

$$(\beta_{opt}^{(\infty)}, s \geq 1) = (b(s), s \geq 1). \qquad (8.76)$$

Prediction for general k can be obtained recursively, by noting that

$$E[X_{n+k}|X_s, s \leq n] = E\{E[X_{n+k}|X_s, s \leq n+j]|X_s, s \leq n\}, \ j \geq 0. \qquad (8.77)$$

Due to the moving average representation (3.12), the mean squared prediction error of $\hat{X}_{n+k}(\beta_{opt}^{(\infty)})$ is equal to

$$MSE_k(\beta_{opt}^{(\infty)}) = \sigma_\epsilon^2 \sum_{j=0}^{k-1} a^2(j). \qquad (8.78)$$

A measure of k-step predictability of the process X_t may be defined as the proportion of σ^2 that is explained by the best prediction $\hat{X}_{n+k}(\beta_{opt}^{(\infty)})$,

$$R_k^2(\beta_{opt}^{(\infty)}) = \frac{\sigma^2 - MSE_k(\beta_{opt}^{(\infty)})}{\sigma^2} = 1 - \frac{MSE_k(\beta_{opt}^{(\infty)})}{\sigma^2}. \qquad (8.79)$$

For illustration, consider ,for instance, a fractional ARIMA(p, d, q) process. The above formulas are particularily simple for this process. The defining equation

$$\phi(B)(1 - B)^d X_t = \psi(B)\epsilon_t \qquad (8.80)$$

can be used directly to obtain the coefficients $a(j)$ and $b(j)$. On one hand, one may write

$$\psi^{-1}(B)\phi(B)(1-B)^d X_t = \epsilon_t \tag{8.81}$$

The coefficients $b(j)$ are then obtained by matching powers in the formal expansion of $\psi^{-1}(B)\phi(B)(1-B)^d$ into a power series. In the same way, the coefficients $a(j)$ are obtained from the representation

$$X_t = (1-B)^{-d}\phi^{-1}(B)\psi(B)\epsilon_t. \tag{8.82}$$

In the case of a fractional $\mathrm{ARIMA}(0,d,0)$ process, we have the explicit formulas (see Proposition 2.2)

$$a(j) = \frac{\Gamma(j+d)}{\Gamma(j+1)\Gamma(d)} \tag{8.83}$$

and

$$b(j) = -\frac{\Gamma(j-d)}{\Gamma(j+1)\Gamma(-d)}. \tag{8.84}$$

Table 8.8 illustrates how much predictions can be improved by using the entire past. For $d = 0.1$ and 0.4 respectively. R_k^2 is calculated for $k = 1, 10, 20,$ and 100. We compare the optimal prediction based on the infinite past with two predictions based on X_n only. The first is the best prediction of X_{n+k} given X_n,

$$E[X_{t+k}|X_t] = \rho(k)X_t. \tag{8.85}$$

The second is

$$\rho^k(1)X_t. \tag{8.86}$$

This is the prediction we would obtain by using an $\mathrm{AR}(1)$ model, in spite of long memory. The average prediction errors are

$$MSE_k(\rho(k)) = \sigma^2(1 - \rho^2(k)) \tag{8.87}$$

and

$$MSE_k(\rho^k(1)) = \sigma^2(1 - 2\rho^k(1)\rho(k) + \rho^{2k}(1)), \tag{8.88}$$

respectively. For comparison, Table 8.9 gives the values of R_k^2 for the case where X_t is an $\mathrm{AR}(1)$ process and the optimal prediction $\hat{X}_{n+k} = \rho^k(1)X_n$ is used. The results show that prediction is much more precise in the presence of long memory, provided that we make proper use of the dependence structure. The size of prediction intervals can be reduced dramatically by using the whole past. Using simple forecasts based on one or only a few observations leads to large losses in precision. Tables 8.8 and 8.9 also

Table 8.8. R_k^2 for three different predictions. The process is a fractional ARIMA$(0, d, 0)$ process with $d = H - \frac{1}{2} = 0.1$ and 0.4 respectively. The predictions are the best linear prediction based on the infinite past, $\rho(k)X_t$ and $\rho^k(1)X_t$.

k	$R_k(\beta_{opt}^{(\infty)})$		$R_k(\rho(k))$		$R_k(\rho^k(1))$	
	$H = 0.6$	$H = 0.9$	$H = 0.6$	$H = 0.9$	$H = 0.6$	$H = 0.9$
1	0.0191	0.4882	0.0123	0.4444	0.0123	0.4444
10	0.0022	0.2706	0.0003	0.1794	10^{-11}	0.0144
20	0.0013	0.2270	0.0001	0.1360	0	0.0002
100	0.0003	0.1475	$7 \cdot 10^{-6}$	0.0714	0	0

Table 8.9. $R_k^2(\beta_{opt}^{(\infty)})$ for an AR(1)-process with $\rho(1) = 1/9$ and $2/3$ respectively. The lag-1 correlation is chosen such that it is the same as for a fractional ARIMA$(0, d, 0)$ process with $H = 0.6$ and 0.9 respectively.

k	$\rho(1) = 1/9$	$\rho(1) = 2/3$
1	0.0123	0.4444
10	$8 \cdot 10^{-20}$	0.0003
20	$7 \cdot 10^{-39}$	$9 \cdot 10^{-8}$
100	0	$6 \cdot 10^{-36}$

illustrate that forecasts of the remote future are much more reliable in the presence of long-range dependence. Theoretically, this can be seen by considering the speed at which the optimal value of R_k^2 converges to zero. For an AR(1) process with $\alpha = \rho(1)$, we have as $k \to \infty$

$$R_k^2 \sim const \cdot \alpha^{2k}. \tag{8.89}$$

Thus, R_k^2 converges to zero exponentially. On the other hand, for a fractional ARIMA$(0, d, 0)$ process,

$$R_k^2 \sim const \cdot \sum_{j=k}^{\infty} j^{2H-3} \sim const \cdot k^{2H-2}. \tag{8.90}$$

Thus, R_k^2 converges to zero hyperbolically. The ratio $\alpha^{2k}/[k^{2H-2}]$ converges to zero for any value of $\alpha \in (-1, 1)$ and $H > 1/2$. The same applies to comparisons between more complicated short-memory and long-memory models.

We apply the results to the Nile River data. To examine the

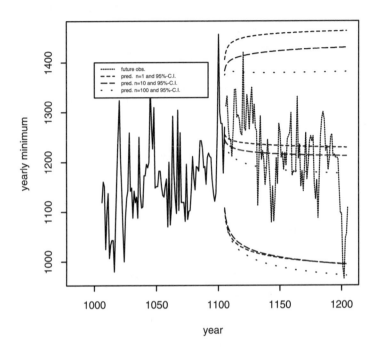

Figure 8.2. *Nile River minima: k-steps-ahead forecasts based on the last n observations* $(n = 1, 10, 100; k = 1, ..., 100)$.

quality of the predictions and prediction intervals, we assume that the observations between the years 1007 and 1106 A.D. are known and compare predictions and 95%-prediction intervals for the next 100 years (1107-1206) with the actually observed values. At first, a fractional ARIMA$(0, d, 0)$ process is fitted to the observed series between 1007 and 1106. The estimates of H and σ_ϵ^2 are 0.963 and 4833, respectively. We compare the forecasts and forecast intervals for three methods: (1) Forecast based on the last observation (year 1106) only, (2) forecast based on the last 10 observations (years 1097 to 1106), and (3) forecast based on all the past 100 observations (years 1007-1106). Except for the first few lags, the predictions based on the last 100 observations are closest to the actually observed "future" values. The level of the Nile River is at a local

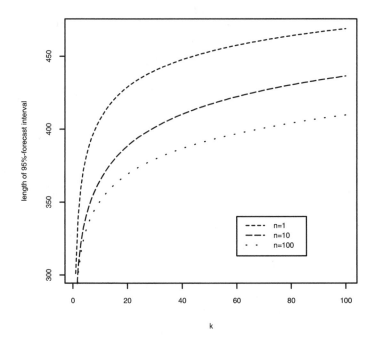

Figure 8.3. *Nile River minima - length of 95%-k-steps-ahead-forecast intervals based on the last n observations ($n = 1, 10, 100; k = 1, ..., 100$).*

maximum around the year 1106. The predictions based on the last 10 and on the last observation are therefore considerably higher. The forecast based on all 100 past observations takes into account that before the current local maximum, the values were much lower than now. Also, Figure 8.3 shows that the 95%-forecast interval is quite a bit shorter for the prediction based on all past observations. Nevertheless, this relatively narrow prediction interval seems to be realistic (see Figure 8.2).

CHAPTER 9

Regression

9.1 Introduction

Suppose that we observe $y = (y_1, ..., y_n)^t$ together with corresponding explanatory variables $x_1 = (x_{1,1}, x_{1,2}, ..., x_{1,p})^t$, $x_2 = (x_{2,1}, x_{2,2}, ..., x_{2,p})^t$, ..., $x_n = (x_{n,1}, x_{n,2}, ..., x_{n,p})^t$. Define the design matrix

$$X = X(n) = n \times p-\text{matrix with rows } x_1^t, x_2^t, ..., x_n^t. \qquad (9.1)$$

We will use the notation

$$x^{(j)} = x^{(j)}(n) = (x_{1,j}, x_{2,j}, ..., x_{n,j})^t \qquad (9.2)$$

for the columns of X. The standard regression model is given by

$$y_t = g(x_t) + \epsilon_t = x_t^t \beta + \epsilon_t, \qquad (9.3)$$

or in matrix notation

$$y = X\beta + \epsilon, \qquad (9.4)$$

with $\epsilon = (\epsilon_1, ..., \epsilon_n)^t$. The explanatory variables $x_{t,j}$ are either fixed or random. The "errors" ϵ_t are identically distributed with expected value zero and variance σ_ϵ^2. Standard assumptions are also that $\epsilon_1, ..., \epsilon_t$ are independent and normally distributed. These assumptions do not always hold exactly for real data. The effect of deviations from normality has been one of the main topics of robust statistics (see, e.g., Huber 1981, Hampel et al. 1986 and references therein). There is also an extended literature on deviations from independence, in the form of autoregressive processes and other short-memory models (see, e.g., Grenander and Rosenblatt 1957, Hannan 1970, Watson 1967, Lomnicki and Zaremba 1957). Here we consider the situation where ϵ_t is a stationary long-memory process. The index t denotes time. We will restrict attention to parametric linear regression (9.3). For some results on simple nonparametric regression see Hall and Hart (1990). Nonparametric density estimation for long-memory processes is discussed briefly in Robinson (1991a).

We will use the following notations. The complex conjugate of a complex number z is denoted by z^*. For two n-dimensional complex valued vectors $a = (a_1, ..., a_n)^t$, $b = (b_1, ..., b_n)^t$, we denote by

$$< a, b >= a^t b^* = \sum_{i=1}^{n} a_i b_i^* \qquad (9.5)$$

the Euclidian scalar product between a and b and by

$$||a|| = \sqrt{< a, a >} \qquad (9.6)$$

the Euclidian norm of a.

As noted earlier, a characteristic feature of long-memory processes is the occurrence of long excursions and local "trends". It should therefore be expected that trends may be difficult to distinguish from the stationary error ϵ_t. One might have to wait quite long before an observed trend can be recognized as real and not just as an accidental increase generated by long memory. A typical example of a situation where the question of trend or no trend gave rise to controversial discussions is the global temperature. The question is whether a "global warming" of the atmosphere took place over about the last 100 years. Consider, for instance, the monthly temperature data for the northern hemishpere introduced in Chapter 1. A thorough discussion of this and related data sets is given, for example in Smith (1993), Bloomfield (1992), and Bloomfield and Nychka (1992). If only this data series is given, the question of global warming amounts to giving a reliable confidence interval for an estimated trend. In first approximation, we may subtract a linear trend. Apart from a slightly larger variability at the beginning, the resulting residuals look approximately stationary (figure 1.11). The periodogram of the residuals (in log-log coordinates) exhibits a clear negative slope, suggesting long memory (Figure 1.12c). To obtain a confidence interval for the linear trend, one therefore needs to derive the distribution of trend estimates under the condition that the residual process ϵ_t has long memory. Theorem 9.1 in the next Section provides this result. It turns out that a polynomial trend is estimated with a precision that is asymptotically infinitely lower than in the iid case. On the other hand, one may ask the question of whether long memory in the residuals always implies that β is more difficult to estimate. The answer is no. The effect of long memory on estimation of the parameters in

(9.3) depends on the structure of the explanatory variables. Consider, for instance, $y = X^t(n)\beta + \epsilon$, where $X(n) = (u_1, ..., u_n)^t$ and u_t $(t = 1, ..., n)$ are independent identically distributed random variables, independent of ϵ_t, $E(u_t) = 0$ and $\sigma_u^2 = \text{var}(u_t) < \infty$. The least squares estimator (LSE) of β is equal to

$$\hat{\beta} = \frac{\sum_{t=1}^{n} y_t u_t}{\sum_{t=1}^{n} u_t^2} = \frac{<y, u>}{||u||^2}.$$

The unconditional variance of $\hat{\beta}$ is approximately equal to

$$n^{-2}\sigma_u^{-2} \sum_{t,s=1}^{n} \gamma(t-s)\delta_{t-s},$$

where $\delta_o = 1$ and $\delta_{t-s} = 0$ for $t \neq s$. Thus $\text{var}(\hat{\beta}) \approx n^{-1}\sigma_\epsilon^2 \sigma_u^{-2}$, which is the same variance as if the errors were independent. An analogous result can be shown for the conditional variance of $\hat{\beta}$ given $u_1, u_2, ..., u_n$.

In the following sections, the interplay between the design matrix X, long memory in the errors, and the distribution of regression estimates is discussed more generally. As in Chapter 8, two main questions need to be answered:

1. What is the asymptotic distribution of the LSE and the BLUE, respectively ?

2. How much do we lose asymptotically, if we use the LSE instead of the BLUE ?

In order that asymptotic limits make sense, some conditions will have to be imposed on the limiting behavior of the design matrix X. We will assume the following general conditions introduced by Grenander (1954):

(1) For all $j = 1, ..., p$,

$$\lim_{n \to \infty} ||x^{(j)}(n)|| = \infty. \tag{9.7}$$

(2) For all $j = 1, ..., p$,

$$\lim_{n \to \infty} \frac{||x^{(j)}(n+1)||}{||x^{(j)}(n)||} = 1. \tag{9.8}$$

(3) For integers $u \geq 0$, define the $n \times 1$ vector

$$x^{(j)}(u, n) = (x_{1+u,j}, x_{2+u,j}, ..., x_{n,j}, 0, ..., 0)^t.$$

Then for each pair j, l and integers $u \geq 0$, the limit

$$R_{j,l}(u) = \lim_{n \to \infty} \frac{< x^{(j)}(n), x^{(l)}(u, n) >}{||x^{(j)}(n)|| ||x^{(l)}(u, n)||} \tag{9.9}$$

exists and is finite. The definition is extended to all integers $u = 0, \pm 1, \pm 2, \dots$ by defining

$$x_{t,j} = 0$$

for $t \leq 0$.

(4) The matrix

$$R(u) = [R_{j,l}(u)]_{j,l=1,\dots,p} \tag{9.10}$$

is nonsingular.

In the case of a random design matrix, (1) to (4) may be postulated in some well defined probabilistic sense, such as, for instance, "almost surely." The first condition makes sure that there is no point t_o such that the contribution of $x_{t,j}, t \geq t_o$ would be negligible. The second conditon makes sure that the absolute value of $x_{t,j}$ does not increase too fast. Condition 3 enables us to define a "correlation" and to use similar techniques as for correlations of a stationary process. The last condition makes sure that the explanatory variables are not linearly dependent in an asymptotic sense.

Examples:

1. polynomial trend:

$$g(x_t) = \beta_o + \beta_1 t + \dots + \beta_q t^q.$$

$$||x^{(j)}(n)||^2 = \sum_{t=1}^{n} t^{2(j-1)} \approx \frac{n^{2j-1}}{2j-1}.$$

Hence, (1) and (2) hold. The elements of the "correlation" matrix R are given by

$$R_{j,l}(u) = \lim_{n \to \infty} \frac{\sum_{t=1}^{n} t^{j-1}(t+u)^{l-1}}{||x^{(j)}(n)|| ||x^{(l)}(u, n)||}$$

$$= \frac{\sqrt{(2j-1)(2l-1)}}{j+l-1}. \tag{9.11}$$

2. Harmonic model: Define

$$g(x_t) = \beta_1 e^{i\lambda_1 t} + \beta_2 e^{i\lambda_2 t} + \dots + \beta_p e^{i\lambda_p t}.$$

Then

$$||x_j(n)||^2 = \sum_{t=1}^{n} e^{i\lambda_j t} e^{-i\lambda_j t} = n.$$

Obviously, conditions (1) and (2) hold. The correlation matrix R is given by

$$R_{jl}(u) = \lim_{n\to\infty} n^{-1} \sum_{t=1}^{n} e^{i(\lambda_j-\lambda_l)t} e^{-i\lambda_l u} = \delta_{jl} e^{-i\lambda_l u}, \qquad (9.12)$$

where $\delta_{jl} = 1$ if $j = l$ and 0 otherwise. The same result is obtained asymptotically for "real versions" of g. For instance, if

$$g(x_t) = \beta_o + \beta_1 \cos(\lambda_1 t) + \ldots + \beta_p \cos(\lambda_p t),$$

then, as $n \to \infty$,

$$||x_j^{(n)}|| \sim n$$

and

$$R_{jl}(u) = \delta_{jl} \cos(\lambda_j t). \qquad (9.13)$$

9.2 Regression with deterministic design

9.2.1 Polynomial trend

Polynomial regression is considered in Yajima (1988). Consider the model

$$y_t = \beta_o + \beta_1 t + \beta_2 t^2 + \ldots + \beta_q t^q + \epsilon_t, \qquad (9.14)$$

where ϵ_t is a stationary process with long memory. In matrix notation, (9.14) can be written as

$$y = X\beta + \epsilon,$$

with the design matrix X given by $X = [x^{(1)}, ..., x^{(q+1)}]$ and

$$x^{(j)} = (1, 2^{j-1}, ..., n^{j-1})^t, \ j = 1, ..., q + 1.$$

Denote by Σ_n the covariance matrix of $\epsilon = (\epsilon_1, ..., \epsilon_n)^t$. The BLUE of β is given by

$$\tilde{\beta} = (X^t \Sigma_n^{-1} X)^{-1} X^t \Sigma_n^{-1} y. \qquad (9.15)$$

Replacing Σ_n by the identity matrix yields the least squares estimator (LSE),

$$\hat{\beta} = (X^t X)^{-1} X^t y. \qquad (9.16)$$

To obtain nondegenerate limit distributions for $\hat{\beta}$ and $\tilde{\beta}$, the following matrices are defined. Let $D(n)$ be the diagonal $(q+1) \times (q+1)$-matrix with diagonal elements

$$D_{jj}(n) = ||x^{(j)}(n)||. \tag{9.17}$$

Define the normalized covariance matrices

$$A_n = D(n)E[(\hat{\beta} - \beta)(\hat{\beta} - \beta)^t]D(n) \tag{9.18}$$

and

$$\tilde{A}_n = D(n)E[(\tilde{\beta} - \beta)(\tilde{\beta} - \beta)^t]D(n). \tag{9.19}$$

Yajima proved the following result for $\hat{\beta}$.

Theorem 9.1 *Suppose that there is a positive continuous function* $f^* : [-\pi, \pi] \to R_+$ *such that the spectral density of* ϵ_t *can be written as*

$$f(\lambda) = f^*(\lambda)|1 - e^{i\lambda}|^{1-2H}, \tag{9.20}$$

with $\frac{1}{2} < H < 1$. *Then the limit*

$$\lim_{n \to \infty} n^{1-2H}A_n \tag{9.21}$$

exists and is equal to

$$A(c_f, H) = 2\pi c_f P^{-1}QP^{-1}, \tag{9.22}$$

where the matrices $P = [p_{jl}]_{j,l=1,...,q+1}$ *and* $Q = Q(H) = [q_{j,l}]_{j,l=1,...,q+1}$ *are defined by*

$$p_{jl} = \frac{\sqrt{(2j-1)(2l-1)}}{j+l-1} \tag{9.23}$$

and

$$q_{jl} = \frac{\sqrt{(2j-1)(2l-1)}\Gamma(2-2H)}{\Gamma(H-\frac{1}{2})\Gamma(\frac{3}{2}-H)} \int_o^1 \int_o^1 x^{j-1}y^{l-1}|x-y|^{2H-2}dxdy, \tag{9.24}$$

In particular, two points are worth noting. The variance of the LSE converges slower to zero than for independent or short-range dependent errors ϵ_t. In the case of short-range dependence, no other normalization is necessary apart from the normalizing matrix $D(n)$. In the case of long memory, the additional normalizing factor n^{1-2H} is needed. The second interesting point is that the asymptotic distribution of $\hat{\beta}$ does not depend on the whole spectral density. It is fully determined by H and the value of f^* at zero.

The shape of the spectrum outside of an arbitrarily small neighborhood of the origin does not matter. This is a generalization of the corresponding result for the sample mean (Theorem 2.2).

For the BLUE, the asymptotic covariance matrix is given by the following theorem in Yajima (1988):

Theorem 9.2 *Under the same assumptions as above, the limit*

$$\lim_{n \to \infty} n^{1-2H} \tilde{A}_n \qquad (9.25)$$

exists and is equal to

$$\tilde{A}(c_f, H) = 2\pi c_f W^{-1}, \qquad (9.26)$$

where the matrix $W = W(H) = [w_{jl}]_{j,l=1,\ldots,q+1}$ *is defined by*

$$w_{jl} = \frac{\sqrt{(2j-1)(2l-1)}}{j+l-2H} \frac{\Gamma(j+\frac{1}{2}-H)\Gamma(l+\frac{1}{2}-H)}{\Gamma(j+1-2H)\Gamma(l+1-2H)}. \qquad (9.27)$$

Again, the rate of convergence of the covariance matrix is slower than under short-range dependence by the factor n^{2H-1} and, the asymptotic covariance matrix depends only on H and $c_f = f^*(0)$.

To calculate the BLUE, the covariances $\gamma(0), \gamma(1), \ldots, \gamma(n-1)$ must be known. The LSE is therefore much more convenient in practice. How much efficiency is lost by using the LSE instead of the BLUE ? The answer depends H and the degree q of the polynomial. A numerical measure for the efficiency of the LSE is the ratio of the determinants of the asymptotic covariance matrices,

$$eff(\hat{\beta}, \tilde{\beta}) = \frac{\det[\tilde{A}(c_f, H)]}{\det[A(c_f, H)]}. \qquad (9.28)$$

In particular, if

$$\lim_{n \to \infty} \tilde{A}_n A_n^{-1} = I, \qquad (9.29)$$

where I is the identity matrix, then this ratio is 1 and the LSE is called *asymtotically efficient*. Note that, for fixed q, (9.28) depends on the value of H only. For $q = 0$, it is equal to the ratio of the two asymptotic variances of the sample mean and the BLUE for the expected value. For comparison, Table 9.1 gives numerical values of $eff(\hat{\beta}, \tilde{\beta})$ for $q = 0$ and 1. It is quite high in both cases. In practice, one will therefore tend to use the LSE instead of the BLUE.

For illustration, consider the temperature data for the northern hemisphere. We fit a linear trend $g(x_t) = \beta_o + \beta_1 t$ to the data and a fractional ARIMA(0,d,0) model to the residuals. The Whittle estimate of H and c_f is 0.87 and 0.0089, respectively. The standard deviation of $\hat{H}$ is 0.028 so that H appears to be clearly above $\frac{1}{2}$.

Table 9.1. *Relative efficiency of the LSE compared to the BLUE as a function of H.*

	$H = 0.5$	$H = 0.7$	$H = 0.9$
$q = 0$	1.000	0.987	0.985
$q = 1$	1.000	0.956	0.901

The least squares estimates of β_o and β_1 are -0.41238 and 0.00032 respectively. To obtain explicit expressions for P [equation (9.23)] and Q [equation (9.24)], note first that for $\alpha < 1$,

$$\int_o^1 \int_o^1 |x - y|^{-\alpha} dx dy = \frac{2}{(1 - \alpha)(2 - \alpha)}$$

$$\int_o^1 x \int_o^1 |x - y|^{-\alpha} dx dy = \frac{1}{(1 - \alpha)(2 - \alpha)}$$

and

$$\int_o^1 \int_o^1 x^2 |x-y|^{-\alpha} dx dy = \frac{1}{(1 - \alpha)(2 - \alpha)}[1 - \frac{2}{3 - \alpha} + \frac{4}{(3 - \alpha)(4 - \alpha)}].$$

Setting $\alpha = 2 - 2\hat{H}$ yields

$$Q = \begin{pmatrix} 1.610 & 1.394 \\ 1.394 & 1.595 \end{pmatrix}.$$

Also,

$$P = \begin{pmatrix} 1 & 0.866 \\ 0.866 & 1 \end{pmatrix},$$

and

$$D(n) = \begin{pmatrix} 40.398 & 0 \\ 0 & 38081.950 \end{pmatrix}.$$

We then obtain

$$A(\hat{c}_f, \hat{H}) = \begin{pmatrix} 0.351 & -0.301 \\ -0.301 & 0.347 \end{pmatrix}.$$

In particular, the estimated standard deviation of $\hat{\beta}$ is equal to

$$n^{\hat{H}-\frac{1}{2}} \frac{\sqrt{A_{22}(\hat{c}_f, \hat{H})}}{D_{22}(n)} = 0.000245.$$

In comparison, the standard deviation obtained under the assumption of independent errors is equal to 0.0000145, which is almost

17 times smaller. The 95%-confidence interval for β_1 is $[0.000293,$
$0.000351]$ under the assumption of independence and $[-0.000158,$
$0.000802]$ if long memory is taken into account. As a consequence
of long-range dependence, we no longer may reject the null hypothesis of no trend. Even a one-sided test of $H_o : \beta_1 = 0$ against
$H_a : \beta_1 > 0$ is not significant. The corresponding P-value is equal
to 0.094. Further refinements may be obtained, for instance, by
taking into account the apparent seasonal component in the residuals (see Figure 1.12c) and the increased variance at the beginning
of the observational period (see Figure 1.11). For instance, to reduce the influence of the seasonal peak in the periodogram on the
estimate of H, we may estimate H from periodogram ordinates
at frequencies that are not too close to $2\pi/12 \approx 0.52$. This can
be done easily in the context of *FEXP* models. Consider, for example, all frequencies that are not in the interval $[0.49, 0.55]$. The
FEXP estimate of H based on the fractional ARIMA$(0,d,0)$ model
is then 0.83. The approximate 95%-confidence interval for β_1 is
$[0.000011, 000633]$. It no longer contains zero, so that we would conclude that β_1 is positive. On the other hand, the values of H and
c_f are estimated. Taking this additional uncertainty into account
would again increase the size of the confidence interval. Overall one
may therefore say that, although there is some indication for global
warming, definite conclusions cannot be drawn based on this data
set only. Taking into account several other data sets can provide
more reliable results (see, e.g., Smith 1993, Bloomfield 1992, and
Bloomfield and Nychka 1992).

9.2.2 General regression with deterministic design

A more general aproach to characterizing the limiting behavior
of slope estimates in regression is to consider the so-called regression spectrum (see, e.g., Priestley 1981, Chapter 7.7). Consider the regression model (9.1) with the sequence of explanatory
real or complex-valued vectors $x_1(n) = (x_{1,1}, ..., x_{1,p})^t$, $x_2(n) =$
$(x_{2,1}, ..., x_{2,p})^t$, $..., x_n(n) = (x_{n,1}, ..., x_{n,p})^t$. Recall that a $p \times p$-
matrix A with complex elements is called Hermitian if

$$A = [A^*]^t.$$

Moreover, A is called positive semidefinite, if $\alpha A \alpha^* \geq 0$ for any
non-zero $p-$dimensional vector α. We will write $A > 0$ if A is a
non-zero positive semidefinite matrix. Grenander and Rosenblatt
(1957) show that, under assumptions (1) to (4) in Section 9.1,

the correlation matrix $R(u)$ is a Hermitian positive semidefinite matrix. This implies that there exists a spectral representation of $R(u)$. More specifically, there exists a matrix valued function $M(\lambda)$ such that for each $\lambda \in [-\pi, \pi]$, $M(\lambda)$ is a Hermitian matrix, the increments

$$\Delta M(\lambda_1, \lambda_2) = M(\lambda_2) - M(\lambda_1) \tag{9.30}$$

are positive semidefinite for any $\lambda_1 < \lambda_2$ and

$$R_{jl}(u) = \int_{-\pi}^{\pi} e^{iu\lambda} dM_{jl}(u) \tag{9.31}$$

The matrix valued function M is called *spectral distribution function* of X. The set $S = \{\lambda \in [-\pi, \pi] : \Delta M(\lambda_1, \lambda_2) > 0 \text{ for all } \lambda_1 < \lambda, \lambda_2 > \lambda\}$ is called the *regression spectrum*. The regression spectrum consists of all points where M increases. In other words, for all $\lambda \in S$,

$$M(\lambda+) - M(\lambda) > 0 \text{ for } \lambda \in S,$$

where

$$M(\lambda+) = \lim_{\nu \downarrow \lambda} M(\nu)$$

denotes the limit from the right.

Examples:

1. Polynomial trend: $R_{jl}(u)$ does not depend on u. In order that

$$R_{jl}(u) = R_{jl}(0) = \int_{-\pi}^{\pi} e^{iu\lambda} dM_{jl}(u)$$

for all u, $dM(\lambda)$ must be zero everywhere, except at the origin. This means that M_{jl} has a jump (point mass) at zero and is constant otherwise. The regression spectrum consists of the point 0 only.

2. Harmonic model: From

$$\delta_{jl} e^{i\lambda_j u} = \int_{-\pi}^{\pi} e^{iu\lambda} dM_{jl}(\lambda)$$

it follows that $M_{jl} \equiv 0$ for $j \neq l$. Moreover, $dM_{jj}(\lambda)$ is not zero for $\lambda = \lambda_j$, and zero otherwise. This means that M_{jj} has a jump at frequency λ_j and is constant otherwise. The regression spectrum consists of the points $\lambda_1, ..., \lambda_p$.

The form of the spectral distribution function M determines the asymptotic distribution of $\hat{\beta}$ and $\tilde{\beta}$. It is characterized by the following theorems. The first two results are due to Grenander and Rosenblatt (1957). They assume that the error process has short memory.

Theorem 9.3 *Assume that* (1) *to* (4) *hold and the errors* ϵ_t *are stationary with a spectral density* f. *Furthermore, suppose that* f *is positive and piecewise continuous in the whole interval* $[-\pi, \pi]$. *Then the following holds.*

(i)

$$\lim_{n \to \infty} A_n = 2\pi R(0)^{-1} \int_{-\pi}^{\pi} f^{-1}(\lambda) dM(\lambda) R(0)^{-1}; \qquad (9.32)$$

(ii)

$$\lim_{n \to \infty} \tilde{A}_n = [\frac{1}{2\pi} \int_{-\pi}^{\pi} f^{-1}(\lambda) dM(\lambda)]^{-1}. \qquad (9.33)$$

This result implies

Theorem 9.4 *Suppose that the assumptions of Theorem 9.3 hold. Then the LSE is asymptotically efficient if and only if* $M(\lambda)$ *increases at no more than p frequencies* $\lambda_1, ..., \lambda_p$ *(and the corresponding symmetric values* $-\lambda_1, ..., -\lambda_p$), *and the sum of the ranks of the increases in* $M(\lambda)$ *is p.*

Examples:

1. Polynomial trend: M_{rs} has a jump (point mass) at zero. The regression spectrum consists of the number zero. The LSE of the polynomial trend is asymptotically efficient.

2. Harmonic model: M has jumps in the diagonal at the frequencies $\lambda_1, ..., \lambda_p$. The LSE is asymptotically efficient.

Both theorems assume the spectral density of the error process ϵ_t to be piecewise continuous. This condition is violated for long-memory processes, as f is infinite at the origin. Using different techniques, Yajima (1991) gave a partial answer to the same questions, for this case. The following notation is needed. Define

$$g_n^{(j)}(\lambda) = \sum_{t=1}^{n} x_{tj} e^{it\lambda},$$

$$m_{jl}^n(\lambda) = \frac{1}{2\pi} \frac{< g_n^{(j)}(\lambda), g_n^{(l)}(\lambda) >}{||x^{(j)}(n)|| ||x^{(l)}(n)||}$$

and

$$M_{jl}^n(\lambda) = \int_{-\pi}^{\lambda} m_{jl}^n(x) \, dx.$$

The $p \times p$-matrix

$$M^n = [M_{jl}^n(\lambda)]_{j,l=1,...,p}$$

is the finite sample version of the spectral distribution function M. Condition (3) implies that the measure $M^n(\lambda)$ (defined on $[-\pi, \pi]$) converges weakly to the measure $M(\lambda)$. This means that, for any continuous function h on $[-\pi, \pi]$,

$$\lim_{n\to\infty} \int_{-\pi}^{\pi} h(\lambda) dM^n(\lambda) = \int_{-\pi}^{\pi} h(\lambda) dM(\lambda).$$

The difference to the situation with a short-memory error process ϵ_t is that the spectral density of ϵ_t has a pole at zero. The key issue is therefore how M^n and M behave at zero. Yajima (1991) considers the following situation:

$$M_{jj}(0+) - M_{jj}(0) > 0, \ 1 \le j \le m, \tag{9.34}$$

where $0 \le m \le p$ and

$$M_{jj}(0+) - M_{jj}(0) = 0, \ j = m+1, ..., p. \tag{9.35}$$

If necessary, the numbering of the explanatory variables has to be changed. If $m = 0$, then (9.34) and (9.35) have to be understood in the sense that none of the diagonal elements M_{jj} has a jump at the origin.

Before stating the results, we define the following matrices. Let $B = [b_{jl}]_{j,l=1,...,m}$ be the $m \times m$-matrix with elements

$$b_{jl} = c_f \lim_{n\to\infty} n^{1-2H} \int_{-\pi}^{\pi} |1 - e^{i\lambda}|^{1-2H} dM_{jl}^n(\lambda). \tag{9.36}$$

Also define the $(p-m) \times (p-m)$-matrix $C = [c_{jl}]_{j,l=1,...,p-m}$ by

$$c_{jl} = \int_{-\pi}^{\pi} f(\lambda) dM_{m+j,m+l}(\lambda) \tag{9.37}$$

and the $p \times p-$matrix

$$V = 2\pi \begin{pmatrix} B & 0 \\ 0 & C \end{pmatrix}. \tag{9.38}$$

Finally, define the $p \times p$ diagonal matrix $D(n, m)$ with diagonal elements

$$D_{jj}(n, m) = D_{jj}(n) n^{H-\frac{1}{2}}, \ j = 1, ..., m \tag{9.39}$$

and

$$D_{jj}(n, m) = D_{jj}(n), \ j = m+1, ..., p, \tag{9.40}$$

where $D(n)$ is defined by (9.17). The distribution of the LSE is given by:

Theorem 9.5 *Suppose that (1) to (4) hold. Then under the assumptions of Theorem 9.1 the following holds.*

(i) Suppose that $m = 0$, i.e., the diagonal elements of M, do not have any jump at zero. Then (9.32) holds, if and only if for any $\delta > 0$ there exists a constant c such that

$$\int_{-c}^{c} f(\lambda) dM_{jj}^{n}(\lambda) < \delta, \ j = 1, ..., p \qquad (9.41)$$

for all sample sizes n.

(ii) Suppose that $m > 0$, (9.41) holds and B exists. Then

$$\lim_{n \to \infty} D^{-1}(n, m)(X^{t}X)E[(\hat{\beta} - \hat{\beta})(\hat{\beta} - \hat{\beta})^{t}](X^{t}X)D^{-1}(n, m) = V.$$

(iii) Suppose that $m > 0$ and (9.41) holds. Then B exists if and only if the integrals

$$\int_{-\pi}^{\pi} |1 - e^{i\lambda}|^{1-2H} dM_{jl} \ (j, l = 1, ..., m)$$

exist.

As the spectral density of ϵ_t has a pole at zero, it is not unexpected that the LSE has the usual rate of convergence only in the case where the regression spectrum does not have a jump at the origin. If zero is an element of the regression spectrum, then it is more difficult to separate the process ϵ_t from the effect of the explanatory variables. A typical example is a polynomial trend. In contrast, the regression spectrum of a seasonal or harmonic component, as defined in the above examples, does not include zero. Therefore, the LSE has the same rate of convergence as for independent errors.

The derivation of the asymptotic distribution of the BLUE is more difficult. Yajima (1991) gave a partial solution. In addition to (1) to (4), he assumed the following conditions:

(5) There exists a $\delta > 2 - 2H$ such that

$$\max_{1 \le t \le n} \frac{x_{tj}^{2}}{||x^{(j)}(n)||^{2}} = o(n^{-\delta}), \ j = 1, ..., p.$$

(6) The jumps of the regression spectrum at zero are limited from above by 1, i.e.,

$$0 < M_{jj}(0+) - M_{jj}(0) < 1, \ 1 \le j \le m.$$

(7) The integral

$$\int_{-\pi}^{\pi} f^{-1}(\lambda) dM(\lambda)$$

is well defined and not equal to zero.

Theorem 9.6 *Under the assumptions of Theorem 9.4 and (1) through (7),*

$$\lim_{n\to\infty} \tilde{A}_n = [\frac{1}{2\pi} \int_{-\pi}^{\pi} f(\lambda)^{-1} dM(\lambda)]^{-1}.$$

If (6) does not hold, then the integral in (7) is no longer well defined. For instance, for polynomial regression, the regression spectrum has jumps $M_{jj}(0+) - M_{jj}(0) = 1$. It was already seen in the previous section that, in addition to the matrix $D(n)$, the normalizing factor n^{1-2H} must be included in order to obtain a finite limiting covariance matrix.

Finally, for the case where the regression spectrum does not include zero, Yajima obtains from the above results conditions under which the LSE is asymptotically efficient.

Theorem 9.7 *Assume that $m = 0$ and (1) through (7) and (9.41) hold. Then the LSE is asymptotically efficient compared to the BLUE, if and only if M increases at not more than p (non-zero) frequencies and the sum of the ranks of the increases is p.*

The results are intuitively plausible, when we compare them with the results of Grenander and Rosenblatt. In Theorems 9.5 to 9.7, the spectral density is continuous outside of an arbitrarily small neighborhood of zero. Therefore, if the regression spectrum contains non-zero frequencies only, one obtains the same result as for spectral densities which are continuous in the whole interval $[-\pi, \pi]$. On the other hand, if zero is an element of the regression spectrum, then the pole of the spectral density changes the efficiency of the LSE.

Unfortunately, the conditions of Theorems 9.5 to 9.7 [in particular condition (9.41)] are not always easy to verify. The following examples are solved explicitly in Yajima (1991):

1. Polynomial trend: The only element of the regression spectrum is zero. The LSE is not asymptotically efficient. The efficiency of the LSE can be obtained directly from Theorems 9.1 and 9.2.

2. Seasonal model $g(x_t) = \beta_1 \cos(\lambda_1 t) + ... + \beta_p \cos(\lambda_p t)$. M has jumps in the diagonal at the frequencies $\lambda_1, ..., \lambda_p$. The LSE is asymptotically efficient. The asymptotic covariance matrix is given by

$$\lim_{n\to\infty} D(n)E[(\hat{\beta} - \beta)(\hat{\beta} - \beta)^t]D(n)$$

$$= \lim_{n\to\infty} D(n)E[(\tilde{\beta} - \beta)(\tilde{\beta} - \beta)^t]D(n)$$

$$= \quad 2\pi \begin{pmatrix} f(\lambda_1) & . & . & 0 \\ 0 & . & . & . \\ . & . & . & . \\ 0 & . & . & f(\lambda_p) \end{pmatrix}$$

9.3 Regression with random design; ANOVA

9.3.1 The ANOVA model

The results in Section 9.2 demonstrate that the asymptotic distribution of the LSE and the BLUE depend on the regression design. Loosely speaking, it appears that the same rate of convergence as under independence is achieved for regressors that change their sign in a "balanced" way. For instance, a seasonal component $\cos \lambda_i t$ changes its sign in a completely symmetric way. In the example of a random explanatory variable u_t discussed in Section 9.1, the expected value of $E(u_t)$ is zero. In this section, we discuss results for more general linear models with random designs. We consider the one-way analysis of variance (ANOVA) model:

$$y_t = \sum_{j=1}^{p} \beta_j x_{t,j} + \epsilon_t, \tag{9.42}$$

where $E(\epsilon_t) = 0$, $\sigma^2 = \text{var}(\epsilon_t) < \infty$, $x_{t,j} \in \{0, 1\}$ and

$$\sum_{j=1}^{p} x_{t,j} = 1. \tag{9.43}$$

The observations y_t are assumed to be taken at the time points $t = 1, ..., n$. In addition to time, t could also stand for any other ordered quantity. For instance, t could be the position on a line in the plane or the distance from a point in the plane, etc. In what follows, the restriction to a scalar index does not appear to be essential. Analogous results are expected to hold, e.g., for spatial data. Also, a generalization to multi-way ANOVA should not be difficult. To simplify the presentation, only the one-way ANOVA model is considered here.

The jth parameter β_j in (9.42) is usually associated with a certain "treatment" number j. Equation (9.43) means that exactly one treatment is assigned at each time point t.

9.3.2 Definition of contrasts

Often one is not interested in knowing the treatment effects β_j themselves. Instead, one would like to compare the effects of the treatments with each other. For instance, one might be interested in individual differences between the effects of two treatment $\beta_j - \beta_k$. More generally, one might be interested in comparing a group of p treatments by defining contrasts

$$\delta(c) = \sum_{j=1}^{p} c_j \beta_j = c^t \beta, \qquad (9.44)$$

where $c = (c_1, ..., c_p)^t$, $\beta = (\beta_1, ..., \beta_p)^t$ and

$$\sum_{j=1}^{p} c_j = 0. \qquad (9.45)$$

Properties of estimates of c follow from the special form of (9.42). Equation (9.42) is a regression model with explanatory variables $x_{t,j}$ which assume only the two values zero and 1. The regression spectrum is fully determined by the sequence in which the treatments are assigned. Usually, treatments are assigned randomly, according to some specified procedure. The main motivation for randomization is to avoid uncontrollable bias. Suppose, for instance, that treatment 1 is assigned a priori to the first k units and treatment 2 to the next $n - k$ units. If for some reason that has nothing to do with the different treatments, the expected value of y_t $(t = 1, ..., k)$ differs from the expected value of y_t $(t = k + 1, ..., n)$, then this effect cannot be distinguished from a possible treatment effect. In the case of dependent errors ϵ_t, and in particular in the case of long-range dependence, there is another problem with this kind of deterministic assignment. Suppose that (9.42) holds with ϵ_t having long-range dependence. The least squares estimate of $\beta_1 - \beta_2$ is the difference of the sample means,

$$k^{-1} \sum_{t=1}^{k} y_t - (n - k)^{-1} \sum_{t=k+1}^{n} y_t$$

$$= \beta_1 - \beta_2 + k^{-1} \sum_{t=1}^{k} \epsilon_t - (n - k)^{-1} \sum_{t=k+1}^{n} \epsilon_t. \qquad (9.46)$$

From the results in Chapter 8 we know that as kn^{-1} tends to a positive constant, the variance of (9.46) is proportional to n^{2H-2}.

Thus, due to the slowly decaying correlations, the contrast $\beta_1 - \beta_2$ is estimated with a much lower precision than under independence or short-range dependence. In contrast, results in Künsch, Beran, and Hampel (1993) show that appropriate random assignment of the treatments results in retaining the same precision as under independence. The main part of these results is summarized in the following sections. It will be sufficient to consider the contrasts

$$\delta_{jk} = \beta_j - \beta_k. \tag{9.47}$$

The variance of estimates of other contrasts can be derived from the results for $\hat{\delta}_{jk}$. The LSE of δ_{jk} is equal to

$$\hat{\delta}_{jk} = \hat{\beta}_j - \hat{\beta}_k, \tag{9.48}$$

where $\hat{\beta}_j, \hat{\beta}_k$, are least squares estimates of β_j and β_k respectively. The BLUE is given by

$$\tilde{\delta}_{jk} = \tilde{\beta}_j - \tilde{\beta}_k, \tag{9.49}$$

where $\tilde{\beta}_j, \tilde{\beta}_k$ are the best linear unbiased estimates of β_j and β_k, respectively.

9.3.3 The conditional variance of contrasts

In the analysis of variance, one is usually interested in conditional inference given the design. That is, inference about contrasts (and other parameters of interest) is done conditionally on $x_{t,j}$ ($t = 1, ..., n$; $j = 1, ..., p$). The values of $x_{t,j}$ are obtained by randomization. However, once the randomization has been done, the design matrix is considered to be fixed. In this sense, the conditional approach leads back to the situation of deterministic designs. The randomization procedure determines which types of design matrices are likely to occur. The approach via probability distributions on design matrices uses the special structure of the explanatory variables in (9.42). The derivation of distributional results can be based on martingale theory (Künsch et al. 1993) and is therefore relatively simple. In contrast, it seems difficult to verify the conditions for deterministic designs given in the previous section, and to combine them with probabilistic statements about the conditional variance.

Before stating the results, we write down the general formulas for the variance of $\hat{\delta}_{jk}$ and $\tilde{\delta}_{jk}$. The conditional variance of $\hat{\delta}_{jk}$ and

$\tilde{\delta}_{jk}$ given the design X is

$$V_n(j,k|X) = \text{var}(\hat{\beta}_j - \hat{\beta}_k|X)$$

$$= \sum_{t,s=1}^{n} \left(\frac{x_{tj}}{n_j} - \frac{x_{tk}}{n_k}\right)\gamma(t-s)\left(\frac{x_{sj}}{n_j} - \frac{x_{sk}}{n_k}\right) \qquad (9.50)$$

and

$$\tilde{V}_n(j,k|X) = \text{var}(\tilde{\beta}_j - \tilde{\beta}_k|X)$$

$$= c^t(X^t\Sigma_n^{-1}X)^{-1}c, \qquad (9.51)$$

where $c_j = 1, c_k = -1$ and $c_i = 0$ for $i \neq j, k$. For uncorrelated errors ϵ_t with variance σ^2, (9.50) and (9.51) simplify to

$$V_n(j,k|X) = \tilde{V}_n(j,k|X) = \sigma^2\left(\frac{1}{n_j} + \frac{1}{n_k}\right). \qquad (9.52)$$

Hence, if $\hat{\sigma}^2$ is an unbiased estimator of σ^2, then

$$\hat{V}_n(j,k|X) = \hat{\sigma}^2\left(\frac{1}{n_j} + \frac{1}{n_k}\right) \qquad (9.53)$$

is an unbiased estimator of both variances.

9.3.4 Three standard randomizations

Künsch, Beran, and Hampel (1993) consider three different randomizations, which are used frequently in practice.

1. *Complete randomization:* Let $a_t \in \{1, ..., p\}$, $t = 1, ..., n$, be a sequence of treatment allocations. This means that, if $a_t = j$, then treatment j is applied to the experimental unit at time t. Define

$$x_{t,j} = 1, \text{ if } a_t = j,$$

and

$$x_{t,j} = 0, \text{ if } a_t \neq j.$$

The allocations a_t are iid with

$$P(a_t = j) = \pi_j. \qquad (9.54)$$

In particular, the number of observations with treatment j,

$$n_j = \sum_{t=1}^{n} x_{tj} \qquad (9.55)$$

is random.

2. *Restricted randomization:* For each treatment j, the number n_j of experimental units to which this treatment is allocated is fixed a priori. All

$$\frac{n!}{n_1! n_2! \cdots n_p!} \tag{9.56}$$

possible allocations $(a_1, ..., a_n)$ with

$$\sum_{t=1}^{n} x_{tj} = n_j \ (j = 1, ..., p) \tag{9.57}$$

are equally likely. Note that, for large samples, restricted randomization is almost the same as complete randomization, if we set

$$\pi_j = \frac{n_j}{n}. \tag{9.58}$$

3. *Blockwise randomization:* For a fixed block length l, divide the time axis into blocks

$$B_k = \{(k-1)l + 1, (k-1)l + 2, ..., kl\} \ (k = 1, ..., b), \tag{9.59}$$

where $b = n/l$. For simplicity, we may assume that b is an integer. The treatment allocations in different blocks are independent from each other. In each block, restricted randomization is applied with

$$\sum_{t \in B_k} x_{tj} = l_j \ (j = 1, ..., p), \tag{9.60}$$

where $\sum l_j = l$.

If the ϵ_ts are uncorrelated, then for the first two randomizations, an unbiased estimator of σ^2 is given by

$$\hat{\sigma}^2 = \frac{1}{n-p} \sum_{t=1}^{n} (y_t - \sum_{j=1}^{p} \hat{\beta}_j x_{tj})^2. \tag{9.61}$$

For blockwise randomization, the estimated block effects

$$\hat{\mu}_k = l^{-1} \sum_{t \in B_k} (y_t - \sum_{j=1}^{p} \hat{\beta}_j x_{tj}) \tag{9.62}$$

are taken into account in the estimation of σ^2. For uncorrelated errors,

$$\hat{\sigma}^2_{block} = \frac{1}{n-p-b+1} \sum_{t=1}^{n} (y_t - \sum_{j=1}^{p} \hat{\beta}_j x_{tj} - \sum_{k=1}^{b} \hat{\mu}_k 1\{t \in B_k\})^2 \tag{9.63}$$

is an unbiased estimator of σ^2.

If the errors have long memory, the conditional variances $V_n(j, k|X)$ and $\tilde{V}_n(j, k|X)$ depend on the correlations $\rho(k)$ ($k = 0, ..., n-1$) [see (9.50) and (9.51)] and are no longer exactly equal to (9.52). Also, in analogy to the results in Section 8.4, the question arises how far $\hat{\sigma}^2$ and $\hat{\sigma}^2_{block}$, and thus (9.53), are biased. In summary, the following questions are addressed in the subsequent sections:

1. How much does $V_n(j, k|X)$ differ from $\sigma^2(n_j^{-1} + n_k^{-1})$?

2. How large is the bias $E[\tilde{V}_n(j, k|X)] - E[V_n(j, k|X)]$?

3. How much does $\tilde{V}_n(j, k|X)$ differ from the asymptotic value of $V_n(j, k|X)$, and thus how efficient is the LSE compared to the BLUE? Are there simple ways of improving the efficiency of the LSE ?

The first two questions refer to the validity (i.e., the accuracy of the level of significance) of standard tests based on the LSE. The third question refers to the power of such tests.

9.3.5 Results for complete randomization

First note that, as $n \to \infty$,

$$\frac{1}{n}\sum_{t=1}^{n} x_{t,j}x_{t+h,k} \to \pi_j\pi_k \qquad (9.64)$$

for $h \neq 0$ and

$$\frac{1}{n}\sum_{t=1}^{n} x_{t,j}x_{t,k} \to \delta_{jk}\pi_j \qquad (9.65)$$

almost surely. If ϵ_t is a stationary process with piecewise continuous spectral density $f = \sigma^2 f_1$ such that $0 < c_1 < f < c_2$ for some constants c_1, c_2, then the asymptotic values of V_n and $\tilde{V}_n$ follow from Grenander and Rosenblatt (1957):

Theorem 9.8 *Under the above conditions the following holds.*

(i) As $n \to \infty$,

$$nV_n(j, k|X) \to \sigma^2(\pi_j^{-1} + \pi_k^{-1}) \qquad (9.66)$$

almost surely.

(ii) As $n \to \infty$,

$$n\tilde{V}_n(j, k|X) \to \sigma^2(\pi_j^{-1} + \pi_k^{-1})[(2\pi)^{-2}\int_{-\pi}^{\pi} f_1(\lambda)^{-1}d\lambda]^{-1} \quad (9.67)$$

almost surely.

The following extension to long-memory processes is derived in Künsch et al. (1993).

Theorem 9.9 *Suppose that the spectral density of ϵ_t is of the form $f = \sigma^2 f_1$ with*

$$f_1(\lambda) = \lambda^{1-2H} f^*(\lambda), \qquad (9.68)$$

where $\frac{1}{2} < H < 1$, f^* is continuous and there exists a constant c such that $0 < c < f^*$. Also assume that f^* is of bounded variation, i.e., there is a constant $\kappa > 0$ such that, for all partitions $0 = \lambda_o < \lambda_1 < ... < \lambda_k = \pi$,

$$\sum_{j=1}^{k} |f^*(\lambda_j) - f^*(\lambda_{j-1})| \leq \kappa.$$

Then the conclusions of Theorem 9.8 hold.

Two important conclusions follow from these results:

1. Inference about contrasts in analysis of variance is usually done under the assumption of independent errors. P-values and confidence intervals provided by standard statistical software packages are calculated under this assumption for the LSE. Theorems 9.8 and 9.9 show that under complete randomization, these P-values and confidence intervals remain to be valid asymptotically, even if the errors are dependent.

2. The asymptotic efficiency of the LSE depends on the shape of the spectral density f via the equation

$$eff(\text{LSE}, \text{BLUE}) = [(2\pi)^{-2} \int_{-\pi}^{\pi} f_1(\lambda)^{-1} d\lambda]^{-1}. \qquad (9.69)$$

For certain dependence structures, (9.69) can be rather small. This leads to a loss of power of the statistical tests. By using the LSE, one therefore tends to discover fewer effects than there may be in the data. For example, if ϵ_t is a fractional $ARIMA(0, H - \frac{1}{2}, 0)$−process with $H = 0.5$, 0.6, 0.7, 0.8, and 0.9, then the asymptotic efficiency (9.69) is equal to 1, 0.97, 0.86, 0.69, and 0.41 respectively.

9.3.6 Restricted randomization

Restricted randomization is slightly more complicated to deal with, since inference is conditional on the p-dimensional vector of integers $(n_1, ..., n_p)^t$. A proof of (9.66) with almost sure convergence replaced by convergence in probability is given in Künschet al. (1993), under the assumption that $n_j n^{-1}$ $(j = 1, ..., p)$ converge to proportions $\pi_j \in (0, 1)$ $(j = 1, ..., p)$. Almost sure convergence and result (9.67) are likely to hold as well, as asymptotically there is no essential difference between full and restricted randomization. Thus, the same remarks on validity and efficiency as above apply here. The order of the bias and the variance of $V_n(j, k|X)$ are given by the following theorem (Künsch et al. 1993):

Theorem 9.10 *Suppose that*

$$\lim_{n \to \infty} \frac{n_j}{n} = \pi_j \in (0, 1) \tag{9.70}$$

for $j = 1, ..., p$. Then, under the same assumptions on ϵ_t as in theorem 9.9, the following holds:

(i)

$$E\{n[V_n - \sigma^2(n_j^{-1} + n_k^{-1})]\} = O(n^{2H-2}) \tag{9.71}$$

(ii)

$$\mathrm{var}(nV_n) = O(c_n) \tag{9.72}$$

where

$$c_n = O(n^{-1}), \text{ if } H < \frac{3}{4},$$

$$c_n = O(n^{-1} \log n), \text{ if } H = \frac{3}{4}$$

and

$$c_n = O(n^{4H-4}), \text{ if } H > \frac{3}{4}.$$

In particular, for $H > \frac{3}{4}$, the standard deviation of V_n is of the same order as the bias. It is also worth noting that the conditions on the error process can be relaxed. Theorem 9.10 holds for any stationary or nonstationary process ϵ_t for which the correlations $\rho(s, t) = \mathrm{corr}(\epsilon_s, \epsilon_t)$ are bounded by

$$|\rho(s, t)| \le C|t - s|^{2H-2}, \tag{9.73}$$

where $0 < C < \infty$ is a constant. This generalization is useful for practical applications, as it might not always be possible to check the assumption of stationarity.

The next question that needs to be answered is what happens if σ^2 is replaced by the estimate (9.61). The exact equation for the expected value of V_n is

$$E[V_n] = (n_j^{-1} + n_k^{-1})\sigma^2[1 - \frac{1}{n(n-1)} \sum_{t \ne s} \rho(t - s)]. \tag{9.74}$$

The conditional expected value of $\hat{\sigma}^2$ is equal to

$$E[\hat{\sigma}^2|X] = \sigma^2(1 - \frac{1}{n-p}\sum_{t \neq s}\rho(t-s)\sum_{j=1}^{p}\frac{x_{tj}x_{sj}}{n_j}). \qquad (9.75)$$

Thus, conditionally on the design X, $(n_j^{-1} + n_k^{-1})\hat{\sigma}^2$ is a biased estimator of V_n. However, putting combining (9.74) and (9.75), we see that

$$E[V_n(j,k|X)] = (\frac{1}{n_j} + \frac{1}{n_k})E[\hat{\sigma}^2]. \qquad (9.76)$$

Thus unconditionally, the bias of $\hat{\sigma}^2$ and the bias of $\sigma^2(n_j^{-1} + n_k^{-1})$ compensate each other exactly. On the average (averaged over designs), $\hat{V}_n$ is unbiased. Asymptotically, the conditional bias also disappears. This follows from Theorem 9.10 and consistency of $\hat{\sigma}^2$. More specifically, the rate of convergence of $\hat{\sigma}^2$ is given by the following result (Künsch et al. 1993):

Theorem 9.11 *Under the same assumptions as above, the following holds.*

(i)

$$\text{var}(E[\hat{\sigma}^2|X]) = O(c_n n^{-2}) \qquad (9.77)$$

where c_n is defined in Theorem 9.1.

(ii) If ϵ_t is Gaussian, then

$$\text{var}(\hat{\sigma}^2|X) = O(c_n) \qquad (9.78)$$

uniformly in X.

For illustration, consider the error process ϵ_t whose covariances are those of a fractional Gaussian noise with $H = 0.5, 0.7$ and 0.9 respectively. Four hundred samples of restricted randomization for $p = 2$ treatments, sample sizes $n = 16$ and $n = 64$, and number of replicates $n_j = \frac{1}{2}n$ were simulated. The average of the simulated exact variances V_n is compared with the corresponding average of the $\hat{V}_n$s. The results in Table 9.2 * illustrate that the biases of $\sigma^2(n_j/n + n_k/n)$ and $\hat{\sigma}^2$ compensate each other.

9.3.7 Blockwise randomization

Randomized blocks are often used in practice when inhomogeneities are expected owing to factors that are not of primary

---

* Adapted from Künsch et al. (1993) with the permission of the author and the Institute of Mathematical Statistics.

Table 9.2. *Ratio $m(V_n)/m(\hat{V}_n)$ for the covariance structure of fractional Gaussian noise, where $m(V_n)$ and $m(\hat{V}_n)$ are simulated average values of V_n and $\hat{V}_n$, respectively.*

	$H = 0.5$	$H = 0.7$	$H = 0.9$
$n = 16$	1.00	1.01	1.01
$n = 64$	1.00	1.01	1.02

interest. Stationary long-memory processes tend to exhibit local though spurious inhomogeneities that can be misinterpreted as non-stationarity. One might therefore suspect that blockwise randomization can improve the accuracy of the LSE for contrasts. Blockwise randomization is very easy to carry out in practice. In contrast, the calculation of the BLUE requires knowledge or estimation of the correlation structure of ϵ_t. It is therefore useful to find simple randomizations where the LSE is almost efficient. For blockwise randomization, the exact formula for the expected value and the variance of V_n is given by the following result (Künsch et al. 1993):

Theorem 9.12 *Under the above assumptions but blockwise randomization with block length l, the following holds:*

(i)

$$E[V_n|X] = \sigma_l^2(n_j^{-1} + n_k^{-1}), \qquad (9.79)$$

where

$$\sigma_l^2 = \sigma^2[1 - \frac{2}{l-1}\sum_{t=1}^{l-1}\rho(t)(1 - \frac{t}{l})] \qquad (9.80)$$

(ii)

$$\text{var}(nV_n) = O(c_n) \qquad (9.81)$$

(iii)

$$nV_n \rightarrow \sigma_l^2(\pi_j^{-1} + \pi_k^{-1}) \qquad (9.82)$$

almost surely, where

$$\pi_j = P(x_{tj} = 1) = l_j/l. \qquad (9.83)$$

Equations (9.79) and (9.80) imply that, if the correlations $\rho(t)$ are strictly decreasing in t and positive, then σ_l^2 is strictly increasing in l (for $l \geq 2$.) Thus, small blocks improve the accuracy of the LSE. Asymptotically, i.e., with l tending to infinity, σ_l^2 converges

Table 9.3. *Efficiency of the LSE for blockwise randomization with $l = 2$
and 4. The error process has the covariance structure of a fractional
ARIMA($0,H - \frac{1}{2},0$) process.*

	$H = 0.5$	$H = 0.7$	$H = 0.9$
$l = 2$	1.00	0.99	0.95
$l = 4$	1.00	0.98	0.91

to σ^2, provided that $\lim_{t \to \infty} \rho(t) = 0$. Finally note that, if σ^2 is replaced by the least squares estimate $\hat{\sigma}^2_{block}$, analogous results hold as above.

How efficient the LSE can become by blockwise randomization is illustrated in Table 9.3[†] The asymptotic efficiency of the LSE is given for $l = 2$ and $l = 4$, where ϵ_t is a fractional ARIMA process with $H = 0.5, 0.7$ and 0.9, respectively. The efficiency is very close to 1 in each of the cases.

[†] Adapted from Künsch et al. (1993) with the permission of the author and the Institute of Mathematical Statistics.

## CHAPTER 10

Goodness of fit tests and related topics

10.1 Goodness of fit tests for the marginal distribution

Let $X_1, ..., X_n$ be a sample of observations from a marginal distribution F. It is often of interest to test the hypothesis that F belongs to a certain class of distributions. For simplicity, we restrict attention to testing normality, i.e., testing the hypothesis

$$H_o : F\left(\frac{x - \mu}{\sigma}\right) = \Phi \qquad (10.1)$$

against

$$H_a : F\left(\frac{x - \mu}{\sigma}\right) \neq \Phi. \qquad (10.2)$$

Here, Φ is the cumulative standard normal distribution, $\mu = E(X_i)$ and $\sigma^2 = \text{var}(X_i)$.

Many different methods for testing (10.1) exist in the literature (for an overview see, e.g., Mardia 1980 and D'Agostino and Stephens 1986). The common feature of most of these methods is that the observations are assumed to be independent. Critical values for test statistics are derived under this assumption. It was noted by Gleser and Moore (1983) and Moore (1982) in the context of short-memory processes that critical values and the corresponding nominal levels of significance can be grossly incorrect, when observations are dependent. To illustrate the effect of long memory, we carry out a small simulation study. Consider the Kolmogorov-Smirnov statistic for the simple hypothesis (μ and σ^2 known) defined by

$$T_{KS}(\mu, \sigma^2) = \sup_x |F_n\left(\frac{x - \mu}{\sigma}\right) - \Phi(x)|, \qquad (10.3)$$

and for the composite hypothesis (μ and σ^2 estimated by $\bar{X}$ and

Table 10.1. *Simulated rejection probabilities for the Kolmogorov-Smirnov test statistic in the simple hypothesis case, with critical values derived under independence for the level of significance 0.05. The results are based on 400 simulated series of fractional Gaussian noise.*

	$H = 0.5$	$H = 0.7$	$H = 0.9$
$n = 20$	0.051	0.219	0.668
$n = 250$	0.059	0.464	0.898

Table 10.2. *Simulated rejection probabilities for the Kolmogorov-Smirnov test statistic in the composite hypothesis case, with critical values derived under independence for the level of significance 0.05. The results are based on 400 simulated series of fractional Gaussian noise.*

	$H = 0.5$	$H = 0.7$	$H = 0.9$
$n = 20$	0.052	0.078	0.061
$n = 250$	0.043	0.106	0.110

s^2, respectively) defined by

$$T_{KS}(\bar{x}, s^2) = \sup_x |F_n(\frac{x - \bar{x}}{s}) - \Phi(x)| \qquad (10.4)$$

where F_n denotes the empirical distribution function. Four hundred series of fractional Gaussian noise of length $n = 20$ and 250 are simulated. For each series, $T_{KS}(\mu, \sigma^2)$ and $T_{KS}(\bar{x}, s^2)$ are calculated. Tables 10.1 and 10.2 [*] give the simulated rejection probabilites with critical regions at the level of significance $\alpha = 0.05$, obtained from quantiles that are valid under independence. We see that the nominal level of significance is clearly incorrect. This is more pronounced the higher the value of H is. Also, there is a clear difference between testing the simple and the composite hypothesis. In the composite hypothesis case, the results are less disastrous than in the simple hypothesis case. A theoretical explanation of these results is given in Beran and Ghosh (1990, 1991). For illustration, we consider the following representative test statistics with $z = (x - \mu)/\sigma$ in the simple hypothesis case and $z = (x - \bar{x})/\sigma$ in

[*] Adapted from Tables 1 and 2 of Jan Beran and Sucharita Ghosh (1991) "Slowly decaying correlations, testing normality, nuisance parameters." JASA, Vol. 86, No. 415, 785-791, with the permission of the authors and the publisher.

the composite hypothesis case:

(i) Kolmogorov-Smirnov (KS) test defined by (10.3) and (10.4) respectively;

(ii) Anderson-Darling statistic

$$T_{AD} = n \int_{-\infty}^{\infty} \frac{(F_n(z) - \Phi(z))^2}{\Phi(z)(1 - \Phi(z))} \phi(z) dz \qquad (10.5)$$

where $\phi(z) = 1/\sqrt{2\pi} \exp(-\frac{1}{2} z^2)$;

(iii) Chi-square statistic

$$T_{\chi^2} = n \sum_{k=1}^{p} \frac{[F_n(l_k) - F_n(l_{k-1}) - (\Phi(l_k) - \Phi(l_{k-1}))]^2}{\Phi(l_k) - \Phi(l_{k-1})}, \qquad (10.6)$$

where $(l_o, l_1], ...(l_{p-1}, l_p]$, is a fixed number p of classes with $-\infty \le l_0 < l_1 < .... < l_p \le \infty$.

(iv) Empirical characteristic function (ecf) (see, e.g., Ghosh 1987, 1993, Ghosh and Ruymgaart 1992)

$$c_n(t) = \frac{1}{n} \sum_{t=1}^{n} e^{itz}, t \in R \qquad (10.7)$$

and its real and imaginary parts, $T_{re}(t)$ and $T_{im}(t)$ respectively.

For the simple hypothesis case, the following results follow from Theorems 3.2 and 3.3, and results by Dehling and Taqqu (1989) on the asymptotic behavior of U-statistics and empirical processes for long-memory processes (see Beran and Ghosh 1991). We denote by $\to_d$ convergence in distribution and by $\to_C$ weak convergence in the supremum norm in $C[-a, a]$ where $a > 0$. Also, c_i $(i = 1, ..., 5)$ are positive constants, and Z a standard normal random variable. χ_1^2 a random variable with χ_1^2-distribution, Z_2 the Rosenblatt process at time $t = 1$, and $\zeta(t)$ a zero mean Gaussian process with covariance function:

$$c(t, s) = E[\zeta(t)\zeta(s)] = \sum_{k=-\infty}^{\infty} [\cosh(ts\gamma(k)) - 1] e^{-\frac{1}{2}(t^2 + s^2)}. \qquad (10.8)$$

Theorem 10.1 *Let X_t be a stationary Gaussian process with spectral density (2.2). Then the following holds:*
(i)

$$n^{1-H} T_{KS} \to_d c_1 |Z| \qquad (10.9)$$

(ii)

$$n^{1-2H} T_{AD} \to_d c_2 \chi_1^2 \qquad (10.10)$$

(iii)

$$n^{1-2H}T_{\chi^2} \to_d c_3\chi_1^2 \qquad (10.11)$$

(iv)

a) If $\frac{1}{2} < H < 1$, then

$$n^{1-H}T_{im}(t) \to_C c_4 \, t \, e^{\frac{1}{2}t^2} Z \qquad (10.12)$$

b) If $\frac{1}{2} < H < 3/4$, then

$$\sqrt{n}[T_{re}(t) - e^{-\frac{1}{2}t^2}] \to_C \zeta(t) \qquad (10.13)$$

c) If $3/4 < H < 1$, then

$$n^{2-2H}[T_{re}(t) - e^{-\frac{1}{2}t^2}] \to_C c_5 t^2 \, e^{-\frac{1}{2}t^2} Z_2. \qquad (10.14)$$

These results have two main implications:

1. For the simple hypothesis, goodness of fit statistics have a slower rate of convergence than under independence. If critical regions obtained under the assumption of independence are used, then the level of significance converges asymptotically to 1 instead of the nominal level α. Thus, long memory has the effect that, even if the null hypothesis is true, it will be rejected for large samples. In order to obtain meaningful significance levels, the statistics have to be normalized by a higher power of n.

2. Theorem 10.1(iv) illustrates that by applying a decompostition into Hermite polynomials, the statistics can be decomposed into an odd and an even part. The odd part has Hermite rank 1, whereas the even part has Hermite rank 2. The even part has a faster rate of convergence, which is even the same as under independence if $\frac{1}{2} < H < \frac{3}{4}$.

The first remark yields an additional explanation to the well-known phenomenon that for large samples, goodness of fit tests always tend to reject the null hypothesis. The usual explanation is that no parametric model is ever correct. Theorem 10.1 shows that the same effect can be caused by long-range dependence.

Analogous limit theorems can be derived for the case of composite hypothesis. The situation turns out to be less dramatic than for the simple hypothesis, as the asymptotic level is not necessarily 1 anymore. Remark 2 above gives the key to understanding the difference from the simple hypothesis case. Note first that the statistics for testing the simple hypothesis are functionals of sums of random variables of the form

$$S = \sum_{j=1}^{n} G\left(\frac{X_j - \mu}{\sigma}, t\right) \qquad (10.15)$$

for a suitably chosen G. The function G can be decomposed into an odd part G_1 and an even part G_2, with Hermite expansions

$$G_1\left(\frac{X - \mu}{\sigma}, t\right) = \sum_{k=1}^{\infty} \frac{a_{2k-1}(t)}{(2k - 1)!} H_{2k-1}\left(\frac{X - \mu}{\sigma}\right) \qquad (10.16)$$

and

$$G_2\left(\frac{X - \mu}{\sigma}\right) = \sum_{k=1}^{\infty} \frac{a_{2k}(t)}{(2k)!} H_{2k}\left(\frac{X - \mu}{\sigma}\right), \qquad (10.17)$$

respectively, with $a_1 \neq 0$ and $a_2 \neq 0$. If μ and σ^2 are replaced by $\bar{X}$ and s^2, then

$$\sum H_1\left(\frac{X_j - \bar{X}}{s}\right) = 0 \qquad (10.18)$$

and

$$n^{-\frac{1}{2}} \sum H_2\left(\frac{X_j - \bar{X}}{s}\right) = -n^{-\frac{1}{2}}. \qquad (10.19)$$

A consequence of (10.18) and (10.19) is that the variance of the test statistics is smaller in the composite hypothesis case. More specifically, Beran and Ghosh (1991) showed that for $\frac{1}{2} < H < 5/6$, the first two terms in the Hermite expansion of (10.15) with estimated μ and σ^2 vanish asymptotically. This implies that, the rate of convergence is not only better than in the case of the simple hypothesis, it is even the same as under independence. Therefore, the levels of the classic tests converge to some $\alpha < 1$, whereas for the fully specified hypothesis they converge to 1.

10.2 Goodness of fit tests for the spectral density

Consider a stationary Gaussian process

$$X_t = \sum_{k=0}^{\infty} a(k; \eta)\epsilon_{t-k} = A(B; \eta)\epsilon_t \qquad (10.20)$$

with iid zero mean normal random variables ϵ_t, $\mathrm{var}(\epsilon_t) = \sigma_\epsilon^2$, and a parametric spectral density

$$f(\lambda; \theta) = \frac{\sigma_\epsilon^2}{2\pi} \left| \sum_{k=0}^{\infty} a(k; \eta)e^{ik\lambda} \right|^2, \qquad (10.21)$$

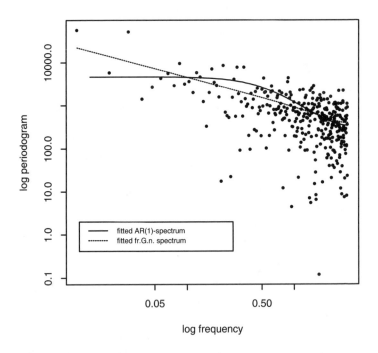

Figure 10.1. *Nile River minima: periodogram and fitted AR(1) and fractional Gaussian noise spectral densities (log-log coordinates).*

where $\theta = (\sigma_\epsilon^2/(2\pi), \eta) \in R^M$ and (2.2) holds for some $H \in (1/2, 1)$. Given a data set $X_1, ..., X_n$, a question one may ask is whether the chosen parametric class of spectral densities $f(\lambda; \theta)$ is appropriate for modeling the observed dependence structure. More specifically, one would like to test the null hypothesis

$$H_o : f(\lambda) \equiv f(\lambda; \theta) \tag{10.22}$$

for some θ, against the alternative

$$H_A : f(\lambda) \not\equiv f(\lambda; \theta) \tag{10.23}$$

for all values of θ.

Example: The log-log periodogram of the Nile River data suggests fitting a straight line to the plot. Mandelbrot (see, e.g., Mandelbrot and Wallis 1968a, 1969a) proposed fractional Gaussian noise. This is a good model candidate as its spectral density in log-log coordinates is almost a straight line and its shape is fully determined by the parameter H only. On the other hand, consider an AR(1) model as an example of a short-memory process with only one parameter. The fitted spectra are displayed in Figure 10.1. The AR(1) spectrum does not seem to capture the negative slope near the origin. However, graphically the deviation from the observed periodogram might not seem much greater than for the fractional Gaussian noise spectrum. The goodness-of-fit test below will answer this question formally.

The following statistic for testing (10.22) is proposed in Beran (1992a). Let

$$A_n(\theta) = \frac{4\pi}{n} \sum_{k=1}^{n^*} [\frac{I(\lambda_j)}{f(\lambda_j; \theta)}]^2 \qquad (10.24)$$

and

$$B_n(\theta) = \frac{4\pi}{n} \sum_{k=1}^{n^*} \frac{I(\lambda_j)}{f(\lambda_j; \theta)}, \qquad (10.25)$$

where $\lambda_j = 2\pi j/n$ $(j = 1, 2, ..., n^*)$ are the Fourier frequencies, n^* is the integer part of $\frac{1}{2}(n-1)$, and $I(\lambda)$ is the periodogram. Note that A_n and B_n are the Riemann sums of the integrals

$$2 \int_o^\pi [\frac{I(\lambda)}{f(\lambda; \theta)}]^2 d\lambda = \int_{-\pi}^\pi [\frac{I(\lambda)}{f(\lambda; \theta)}]^2 d\lambda \qquad (10.26)$$

and

$$2 \int_o^\pi \frac{I(\lambda)}{f(\lambda; \theta)} d\lambda = \int_{-\pi}^\pi \frac{I(\lambda)}{f(\lambda; \theta)} d\lambda, \qquad (10.27)$$

respectively. Also, let $\hat{\theta}$ be Whittle's approximate MLE. Note that in terms of the notation here this means that $\hat{\eta}$ minimizes $B_n(\theta^*)$ with respect to η, where $\theta^* = (1, \eta)$, and the scale parameter is estimated by $\hat{\theta}_1 = \hat{\sigma}_\epsilon^2/(2\pi) = (2\pi)^{-1} B_n(\hat{\theta}^*)$. We then define the test statistic

$$T_n(\hat{\theta}) = \frac{A_n(\hat{\theta})}{B_n^2(\hat{\theta})}. \qquad (10.28)$$

In the case of short-memory processes, this statistic was proposed by Milhoj (1981). Intuitively, T_n is a standardized overall measure of the discrepancy between the periodogram and the fitted spectrum. The discrepancy at a specific frequency λ is defined to be proportional to $[I(\lambda)/f(\lambda; \hat{\theta})]^2$. An interpretation in the time domain is obtained in terms of the correlations of the estimated residual process

$$\epsilon_t(\hat{\eta}) = A^{-1}(B; \hat{\eta})X_t. \tag{10.29}$$

First note that the autocovariances of $\epsilon_t(\hat{\eta})$ can be estimated by

$$\hat{\gamma}_\epsilon(k) = \frac{4\pi}{n} \sum_{j=1}^{n^*} \left[\frac{I(\omega_j)}{f(\omega_j; \hat{\theta}^*)}\right] \cos(k\omega_j). \tag{10.30}$$

The autocorrelations of the residual process are estimated by

$$\hat{\rho}_\epsilon(k) = \frac{\hat{\gamma}_\epsilon(k)}{\hat{\gamma}_\epsilon(0)}. \tag{10.31}$$

Then T_n can be written as

$$T_n = \frac{1}{2\pi} \sum_{k=0}^{n-1} \hat{\rho}_\epsilon^2(k). \tag{10.32}$$

Thus, T_n is the sum of the squares of all estimable correlations of the residual process obtained by fitting the chosen model. If the parametric model is correct, then the residual process is uncorrelated, otherwise not all correlations are zero. T_n is an extension of the portmanteau statistic based on a finite number of correlations of $\epsilon_t(\eta)$ (see e.g. Box and Pierce 1970; Davies, Triggs and Newbold 1977; Box and Ljung 1978). The portmanteau statistic is of the form

$$T_{p;n} = \frac{1}{2\pi} \sum_{k=0}^{p} \hat{\rho}_k^2, \tag{10.33}$$

where p is a fixed integer. Instead of restricting attention to a fixed set of correlations, (10.32) takes into account all estimable correlations simultaneously. On one hand, this is a disadvantage. The correlations at lags close to n are estimated very poorly. Thus, including such correlations leads to more variability of T_n, even if H_o is true. On the other hand, in contrast to (10.33), T_n is a consistent test. Deviations at any arbitrary lag can be detected. Including all correlations is particularly appropriate for long-memory spectra. Suppose, for instance, that a slightly incorrect model leads to underestimation of H. The residual process will then be a long-memory process. It was noted previously that single autocorrelations of such a process need not be particularly large. It is rather the decay of the correlations that is unusually slow. T_n will then tend to be more powerful than test statistics based on a small fixed number of correlations.

The distribution of T_n under H_o follows from the following result (see Milhoj 1981 for short-memory processes and Beran 1992a for long-memory processes).

Theorem 10.2 *Define* $\mu = (\mu_1, \mu_2)$ *by* $\mu_1 = 4\pi$, $\mu_2 = 2\pi$ *and* $\Omega = (\omega_{jl})_{j,l=1,2}$ *by* $\omega_{11} = 160\pi^2$, $\omega_{12} = \omega_{21} = 32\pi^2$ *and* $\omega_{22} = 8\pi^2$. *Then, under* H_o, *the following holds.*

(i)

$$E[A_n(\hat{\theta})] = \mu_1 + o(n^{-1/2}) \tag{10.34}$$

$$E[B_n(\hat{\theta})] = \mu_2 + o(n^{-1/2}) \tag{10.35}$$

(ii)

$$\sqrt{n}[A_n(\hat{\theta}) - \mu_1, B_n(\hat{\theta}) - \mu_2] \to_d Z, \tag{10.36}$$

where Z *is a bivariate Gaussian random variable with zero mean and covariance matrix* Ω.

In particular, it follows that the distribution of T_n does not depend on the unknown parameter θ and is the same, as if θ were known (simple hypothesis). An analogous limit theorem can be obtained under certain types of alternatives (Milhoj 1981 and Beran 1992a). Critical values of T_n are obtained from

$$P(T_n \le c) = P(A_n(\theta)/B_n^2(\theta) \le c) = P(A_n(\theta) - c \, B_n^2(\theta) \le 0)$$

$$\approx \int\int_{x \le cy^2} \phi_n(x, y) dx \, dy, \tag{10.37}$$

where ϕ_n is the bivariate normal distribution with mean μ and covariance matrix $n^{-1}\Omega$. A further simplification can be achieved by noting that T_n itself is asymptotically normal with mean π^{-1} and variance $2\pi^{-2}n^{-1}$. Hence

$$P(T_n \le c) \approx \Phi(\sqrt{\frac{n}{2}}(c\pi - 1)). \tag{10.38}$$

Numerical comparisons in Milhoj (1981) show that (10.38) is close to (10.37) for moderately large sample sizes (for example, $n = 128$).

Example: The value of T_n for the Nile River data and fractional Gaussian noise as parametric model is 0.309. The data consist of 660 measurements. The approximate P-value is $P(T_n \ge 0.309) \approx 1 - \Phi(-0.53) = 0.7019$. On the other hand, for the AR(1) model, $T_n = 0.420$ with a P-value of $1 - \Phi(5.80) \approx 0$. Thus, fractional Gaussian noise appears to model the spectral shape very well, whereas the AR(1) spectrum clearly deviates from the observed spectrum. Finally note that a test of independence yeids $T_n = 2.081$ with a P-value of $1 - \Phi(100.60) = 0$!

A more specific question than the hypothesis (10.22) is whether or not H is larger than $\frac{1}{2}$. In the spectral domain this means that we are only interested in how to model the spectral density near the origin. A modification of the above test to testing H_o for a certain range of frequencies $0 < \lambda < \gamma < \pi$ is given in Beran (1992a). As an alternative, one may want to test directly the hypothesis

$$H_o : H = \frac{1}{2} \tag{10.39}$$

against the alternative

$$H_a : H > \frac{1}{2}. \tag{10.40}$$

Suppose that H is estimated by $\hat{H}$ and that

$$Z_n = \frac{\hat{H} - H}{\sqrt{\mathrm{var}(\hat{H})}} \tag{10.41}$$

is asymptotically standard normal. This is the case for the estimators discussed Chapters 5-7 and Section 4.6. Under this assumption, a critical region at the level of significance α is given by $z > z_\alpha$ where $\Phi(z_\alpha) = 1 - \alpha$.

Another procedure for testing (10.39) against (10.40) is given in Davies and Harte (1987) for the case of fractional Gaussian noise. They derive so-called locally optimal and beta-optimal tests. A test is called locally optimal if it maximizes the derivative of the power function with respect to H at $H = \frac{1}{2}$. The proposed tests assume however the specific correlation form of fractional Gaussian noise. They therefore seem to be of limited practical interest.

10.3 Changes in the spectral domain

The periodogram of the Nile River data is very well approximated by the spectral density of fractional Gaussian noise. However, a striking feature of the plot of the observations against time (Figure 1.3) is that about the first 100 observations seem to fluctuate much more independently than the subsequent 500 measurements. Whittle's MLE of H is equal to 0.54 and 0.88 for the first 100 and the next 500 observations, respectively. The natural question is: Can this discrepancy be explained by randomness that is due to estimation or does H actually change after the first 100 observations? The answer is not obvious a priori, because the estimate of H is rather uncertain when it is based on a short series only.

A partial answer is given by the following result which is a corollary of Theorem 3.6 (Beran and Terrin 1994):

Theorem 10.3 *Let X_t be a linear process* (3.12) *with spectral density* (2.2). *Let $\hat{\eta}^{(1)}, ..., \hat{\eta}^{(k)}$ be the Whittle estimates of η based on the subseries* $(X_{n_o}, ..., X_{n_1}), (X_{n_1+1}, ..., X_{n_2}), ..., (X_{n_{k-1}}, ..., X_{n_k})$, *respectively, where $n_o = 1 < n_1 < ... < n_k = n$. Define the matrix*

$$\hat{E} = [\hat{\eta}^{(1)} - \eta, ..., \hat{\eta}^{(k)} - \eta]. \tag{10.42}$$

Suppose that there exist $0 < \gamma_l < 1$ $(l = 1, ..., k-1)$ such $n_l/n \rightarrow \gamma_l$ $(l = 1, ..., k - 1)$ as $n \rightarrow \infty$. Then

$$\sqrt{n}\hat{E} \rightarrow_d E, \tag{10.43}$$

where E is a zero mean matrix with independent columns. Each column l $(l = 1, ..., k)$ is asymptotically normal with covariance matrix

$$C(\gamma_l; \eta) = \gamma_l^{-1}C(\eta), \tag{10.44}$$

where $C(\eta)$ is defined by (5.25).

Consider now testing the hypothesis that X_t is stationary with a constant value of H against the alternative that H changes at some time points $1 < n_1 < ... < n_{k-1} < n$. Denote by $v(\hat{\eta})$ the asymptotic variance of $\sqrt{n}(\hat{H} - H)$, evaluated at the estimated parameter value $\hat{\eta}$ and let $\bar{H}$ be the sample mean of $\hat{H}^{(1)}, ..., \hat{H}^{(k)}$. If $n_1, ..., n_{k-1}$ are known a priori, then a simple test statistic is given by

$$T_{1,2,...,k} = \sum_{j=1}^{k} n_j \frac{(\hat{H}^{(j)} - \bar{H})^2}{v(\hat{\eta}^{(j)})}. \tag{10.45}$$

By the above theorem, an asymptotically correct rejection region at the level of significance α is given by $T_{1,2,...,k} > \chi^2_{k-1;\alpha}$ where $\chi^2_{k-1;\alpha}$ is the upper $(1 - \alpha)$-quantile of the χ^2-distribution with $k - 1$ degrees of freedom.

Applying this to the Nile River data, we obtain the following result. For simplicity, we consider the first 600 observations and divide them into 6 series, each of length 100. The estimates of H for the six subseries are 0.54, 0.85, 0.86, 0.83, 0.84, and 0.93, respectively. The variability of $\hat{H}^{(j)}$ seems high; however, the estimates of H are based on 100 observations only so that, in principle, high variability has to be expected even if H is constant. The biggest jump occurs between $j = 1$ and $j = 2$. The test statistic $T_{1,2,3,4,5,6}$

is equal to

$$T_{1,2,3,4,5,6} = \sum_{j=1}^{6} 100 \frac{(\hat{H}^{(j)} - \bar{H})^2}{v(\hat{H}^{(j)})}$$

$$= 18.187 + 0.389 + 0.596 + 0.106 + 0.225 + 3.260 = 22.762.$$

The P-value is

$$P(T_{1,2,3,4,5,6} > 22.762) \approx P(\chi_5^2 > 22.762) = 0.0004.$$

Hence, there appears to be statistical evidence that H is not constant. Clearly, the first term causes rejection of H_o. This corresponds to the visual impression of the time series plot. It should be noted, however, that Theorem 10.3 assumes that the time points where H might change are known a priori. Instead, we first looked at the time series plot and then decided a posteriori that k might be equal to 2 and n_1 is about 100. The given formal level of significance obtained from Theorem 10.3 is therefore not quite correct. In order to incorporate lack of knowledge about $n_1, ..., n_{k-1}$, one would need to use a statistic that compares $\hat{H}$ for many different subseries simultaneously. A nonparametric method that makes such comparisons for the whole spectral density is given in Giraitis and Leipus (1991, 1992). It is an extension of results derived by Picard (1985) for short-memory processes (see also Epps 1988). The method can be described as follows. Suppose that X_t is a linear process (3.12), all moments of ϵ_t are finite and the spectral density f is of the form (2.2). Consider the null hypothesis that f is the same for the whole series against the alternative that there is a time point n_1 after which the spectral density is different. For $0 \leq m \leq n$, denote by

$$I_m^{(1)}(\lambda) = \frac{1}{2\pi m} |\sum_{t=1}^{m} e^{it\lambda} X_t|^2 \tag{10.46}$$

the periodogram of $X_1, ..., X_m$ and by

$$I_m^{(2)}(\lambda) = \frac{1}{2\pi(n-m)} |\sum_{t=m+1}^{n} e^{it\lambda} X_t|^2 \tag{10.47}$$

the periodogram of $X_{m+1}, ..., X_n$. Also, let

$$\hat{F}_m^{(1)}(\lambda) = \int_o^{\lambda} I_m^{(1)}(x) dx \tag{10.48}$$

and

$$\hat{F}_{n-m}^{(2)}(\lambda) = \int_o^{\lambda} I_m^{(2)}(x) dx \tag{10.49}$$

be the corresponding cumulative periodograms. For $(t, \lambda) \in K = [0, 1] \times [0, \pi]$, define

$$\zeta_n(t, \lambda) = \frac{[nt]}{n}(1 - \frac{[nt]}{n})[\hat{F}^{(1)}_{[nt]}(\lambda) - \hat{F}^{(2)}_{n-[nt]}(\lambda)]. \qquad (10.50)$$

Picard (1985) proposes the test statistic

$$T_n = \sqrt{n} \sup_{(t,\lambda)\in K} |\zeta_n(t, \lambda)|. \qquad (10.51)$$

This test statistic measures the largest discrepancy between the cumulative periodograms that one can achieve by splitting the series into two halves. It can be shown that, under the alternative,

$$\lim_{n\to\infty} P(T_n > c) = 0 \qquad (10.52)$$

for any $c > 0$. This means that a test based on the rejection region $T_n > c$ is consistent. Moreover, Giraitis and Leipus obtained a formula for the asymptotic tail distribution of T_n under the null hypothesis.

Theorem 10.4 *Let χ_4 be the fourth cumulant of ϵ_t,*

$$F(\lambda) = \int_o^\lambda f(x)dx, \qquad (10.53)$$

$$\sigma(\lambda, \nu) = 2\pi \int_o^{\min(\lambda,\nu)} f^2(\lambda)d\lambda + \chi_4 F(\lambda)F(\nu) \qquad (10.54)$$

$$\tau(t, \lambda; s, \nu) = [\min(t, s) - ts] \cdot \sigma(\lambda, \nu) \qquad (10.55)$$

where $(t, \lambda), (s, \nu) \in K$. Suppose that the null hypothesis is true. Then, under the above conditions and some additional regularity assumptions,

$$\lim_{n\to\infty} P(T_n > c) = P(\sup_{(t,\lambda)\in K} |Z(t, \lambda)| > c), \qquad (10.56)$$

where $Z(t, \lambda)$ is a Gaussian process with two-dimensional index (t, λ), zero mean, and covariance function τ.

Based on this result, Giraitis and Leipus also derived a method for estimating the time point where the change in the spectral density occurs and an estimator for the maximal jump in the cumulative spectrum,

$$\sup_{\lambda\in[0,\pi]} |F_1(\lambda) - F_2(\lambda)| \qquad (10.57)$$

where F_1, F_2 are the cumulative spectral densities for the process before and after the change point, respectively.

To conclude this section, we note another application of Theorem 10.3. In some situations, H has to be estimated for a very long time series. This occurs in particular in technical applications. For instance, the VBR and Ethernet data introduced in Chapter 1 are parts of series of several tens of thousands of measurements. In these applications, one often needs quick (sometimes "real time") estimation of H (and possibly other parameters $\eta_2, ..., \eta_M$). Theorem 10.3 implies that, if the series is divided into sufficiently long subseries, then the estimates $\hat{\eta}_1, ..., \hat{\eta}_k$ based on the different subseries are approximately independent. Thus, instead of estimating η from the whole series, one can divide the series into subseries $(X_1, ..., X_j), (X_{j+1}, ..., X_{2j}), ..., (X_{(k-1)j+1}, ..., X_{kj})$, $(j > 0, k = [n/j])$. The parameter estimate $\hat{\eta}$ may then be defined by

$$\bar{\eta} = k^{-1} \sum_{j=1}^{k} \hat{\eta}_j. \tag{10.58}$$

By Theorem 10.3, as $n \to \infty$ and $jn^{-1} \to \gamma > 0$, $\sqrt{n}(\hat{\eta} - \eta)$ has the same asymptotic distribution as the Whittle estimator based on the whole series. For Gaussian processes $\bar{\eta}$ is asymptotically efficient. Calculation of (10.46) is particularly fast on computers with several parallel processors. Moreover, as a byproduct, splitting the series provides the possibility of checking whether the parameters remain constant.

## CHAPTER 11

Miscellaneous topics

11.1 Processes with infinite variance

In all previous chapters, a basic assumption was that the variance of X_t is finite. In some areas of application this might be an unduly restrictive assumption. For example, Mandelbrot and Wallis (1968a) observed that models with infinite variance are more appropriate for many hydrological series. In analogy to the "Joseph effect," they call the phenomenon of an infinite variance the "Noah effect." Other examples where processes with an infinite variance might be more appropriate are reported, for instance, in Stuck and Kleiner (1974; telephone signals), Fama (1965; stock market prices) and Mandelbrot (1969; economic data sets). For more references, see, for instance, Davis and Resnick (1986a,b). Generally speaking, infinite variance processes are good model candidates for "bursty" phenomena, i.e., phenomena that exhibit occasional unusually large observations.

There is an extensive literature on the probability theory for infinite variance processes (for references see Samorodnitsky and Taqqu 1993). Typical examples are linear processes (see Brockwell and Davis 1987, Chapter 12.5), stable self-similar processes (see, e.g., Kono and Maejima 1990, Cambanis and Maejima 1989, Cambanis et al. 1992, Samorodnitsky and Taqqu 1989, 1990, 1992), and ARIMA and fractional ARIMA processes with infinite variance innovations. A model that may be especially useful for statistical inference is the fractional ARIMA model with stable innovations discussed in Kokoszka and Taqqu (1993c). It is defined by

$$\phi(B)(1 - B)^d X_t = \psi(B)\epsilon_t, \tag{11.1}$$

where the innovations ϵ_t are iid stable random variables. The case $d = 0$ and related linear processes with infinite variance are also considered, for instance, in Bhansali (1984, 1988), Hannan and Kanter (1977), Klüppelberg and Mikosh (1991, 1992), Gadrich and

Adler (1993) (see also references therein). The purpose of the extension to $d > 0$ is to obtain processes that have "long memory" in some sense. However, before starting to talk about "short" and "long" memory, one needs to define how dependence is measured when second moments do not exist. Autocovariances are no longer defined.

Several measures of dependence are proposed in the literature. For instance, a direct extension of the usual autocorrelation is obtained by considering the sample correlations

$$\hat{\rho}(k) = \frac{\sum_{t=1}^{n-|k|} X_t X_{t+|k|}}{\sum_{t=1}^{n} X_t^2}. \tag{11.2}$$

Because observations consist of finite numbers that are almost surely never all zero, (11.27) is always defined. Davis and Resnik (1986) consider linear processes

$$X_t = \sum_{j=-\infty}^{\infty} \psi_j \epsilon_{t-j} \tag{11.3}$$

with iid stable innovations ϵ_s, and prove that under certain summability conditions on the coefficients ψ_j, $\hat{\rho}(k)$ converges almost surely to

$$\rho(k) = \frac{\sum_{j=-\infty}^{\infty} \psi_j \psi_{j+k}}{\sum_{j=-\infty}^{\infty} \psi_j^2}. \tag{11.4}$$

Also, a limit theorem for the properly normalized variables $\hat{\rho}(k)$ is obtained. In the case where the variance of ϵ_s is finite with $E(X_t) = 0$, $\rho(k)$ coincides with the usual autocorrelation. Another measure of dependence is the so-called codifference:

$$\tau(u) = \log E[e^{i(X_{t+u} - X_t)}] - \log E[e^{iX_{t+u}}] - \log E[e^{iX_t}]. \tag{11.5}$$

For a Gaussian process, $\tau(u)$ coincides with the autocovariance $\gamma(u)$. Kokoszka and Taqqu (1993a) showed that under suitable conditions, the codifference of the fractional ARIMA process (11.27) has a hyperbolic decay. This allows for the distinction between long and short memory in terms of the summability of τ. For other measures of dependence, more detailed results, and further references, we refer the interested reader, for instance, to Kokoszka and Taqqu (1992, 1993a,b,c).

11.2 Fractional GARMA processes

Long memory is defined in terms of the spectral density by assuming that f has a pole at the origin. Gray, Zhang, and Woodward (1989) generalize this definition by allowing the spectral density to have poles at any (zero or non-zero) frequency $\lambda_o \in [0, \pi]$. They say that a stationary process X_t has *long memory* if f has a pole at some frequency $\lambda_o \in [0, \pi]$. If $\lambda_o \neq 0$, then the correlations are summable, because they converge to zero and change their sign periodically. However, the absolute values of the correlations are not summable (see Theorem 11.2 below). The motivation for considering poles outside of the origin is to model persistent cyclic behavior. A pole of the spectrum at a frequency $\lambda_o \neq 0$ implies that the process has a long-lasting nondeterministic periodic component. More specifically, Gray et al. consider the following generalization of a fractional ARIMA process.

Definition 11.1 *Let ϵ_t, $t \in Z$ be a sequence of independent zero mean random variables with variance σ_ϵ^2. Define X_t by*

$$X_t = (1 - 2uB + B^2)^{-\frac{d}{2}} \epsilon_t \tag{11.6}$$

for some $|u| \leq 1, d \neq 0$. Then X_t is called a Gegenbauer process with parameters u and $\lambda = \frac{1}{2}d$.

In particular, for $u = 1$ one obtains the fractional ARIMA(0,d,0) process. The coefficients $a(j; u)$ in the infinite moving average representation

$$X_t = (1 - 2uB + B^2)^{-\frac{d}{2}} \epsilon_t = \sum_{j=0}^{\infty} a(j; u)\epsilon_{t-j} \tag{11.7}$$

are so-called Gegenbauer polynomials. They are of the form

$$a(j; u) = \sum_{k=0}^{[\frac{j}{2}]} (-1)^k \frac{\Gamma(\frac{d}{2} + j)}{\Gamma(\frac{d}{2})} \frac{(2u)^{j-2k}}{k!(j - 2k)!}. \tag{11.8}$$

Stationarity and invertibility of X_t are established by the following result (Gray et al. 1989):

Theorem 11.1 *Let X_t be defined by (11.36). Then*

(i) X_t is stationary if either

$$|u| < 1 \text{ and } d < 1$$

or

$$u = \pm 1 \text{ and } d < \frac{1}{2};$$

(ii) X_t is invertible, if either

$$|u| < 1 \text{ and } d > -1$$

or

$$u = \pm 1 \text{ and } d > -\frac{1}{2}.$$

In the stationary region, the spectral density of X_t is given by

$$f(\lambda) = \frac{\sigma_\epsilon^2}{2\pi} 2^{-d}(\cos \lambda - u)^{-d}. \tag{11.9}$$

For $\lambda_o = \arccos(u)$, f is infinite. In the case of a fractional ARIMA$(0,d,0)$ process we have $u = 1$ so that $\lambda_o = 0$. The correlations of X_t are given by

Theorem 11.2   *(i) For $u = 1$ and $0 < d < \frac{1}{2}$,*

$$\rho(k) = \frac{\Gamma(1-d)}{\Gamma(d)} \frac{\Gamma(k+d)}{\Gamma(k+1-d)}. \tag{11.10}$$

In particular, for $k \to \infty$,

$$\rho(k) \sim_{k \to \infty} \text{const} \cdot k^{2d-1}. \tag{11.11}$$

(ii) For $u = -1$ and $0 < d < \frac{1}{2}$,

$$\rho(k) = (-1)^k \frac{\Gamma(1-d)}{\Gamma(d)} \frac{\Gamma(k+d)}{\Gamma(k+1-d)}. \tag{11.12}$$

In particular, for $k \to \infty$,

$$\rho(k) \sim_{k \to \infty} \text{const} \cdot (-1)^k k^{2d-1} \tag{11.13}$$

(iii) For $|u| < 1$ and $0 < d < 1$,

$$\rho(k) \sim_{k \to \infty} \text{const} \cdot k^{d-1} \sin(\frac{\pi}{2}d - k\lambda_o) \tag{11.14}$$

as $k \to \infty$.

Case (i) is long memory in the sense of the original definition in Chapter 2. Case (iii) is a new kind of long memory. The correlations decay slowly in the sense that the sharp asymptotic upper bound

$$|\rho(k)| \leq \text{const} \cdot k^{2d-1} \tag{11.15}$$

converges slowly to zero. For large lags, the sign and the size of individual correlations change periodically.

Equation (11.7) can be generalized to include autoregressive and moving average components. In analogy to a fractional

ARIMA(p, d, q) process, a GARMA(p, d, q) process is defined by

$$\phi(B)(1 - 2uB + B^2)^{\frac{d}{2}} X_t = \psi(B)\epsilon_t, \qquad (11.16)$$

where ϕ and ψ are suitably chosen polynomials. The asymptotic behavior of the correlations is essentially the same as above. The spectral density is given by

$$f(\lambda) = \frac{\sigma_\epsilon^2}{2\pi} \frac{|\phi(e^{i\lambda})|^2}{|\psi(e^{i\lambda})|^2} 2^{-d} (\cos \lambda - u)^{-d}. \qquad (11.17)$$

11.3 Simulation of long-memory processes

11.3.1 Introduction

Many methods exist for simulating Gaussian processes. The most obvious method is to multiply a vector of iid standard normal random variables by a suitable transformation matrix. More specifically, suppose that we want to simulate n observations of a stationary Gaussian process with zero mean and given autocovariances $\gamma(k)$. This means that a vector $X = (X_1, ..., X_n)^t$ has to be drawn randomly from an n-dimensional normal distribution with mean 0 and covariance matrix

$$\Sigma_n = [\gamma(i - j)]_{i,j=1,...,n} \qquad (11.18)$$

This can be achieved by the following steps:

1. Write the covariance matrix as a product of a real lower triangular matrix L and its transposed L^t,

$$\Sigma = LL^t. \qquad (11.19)$$

This representation is called Cholesky decomposition. For a given model, this step has to be performed only once.

2. Simulate a vector of independent standard normal random variables

$$Z = (Z_1, ..., Z_n)^t. \qquad (11.20)$$

3. Define

$$X = (X_1, ..., X_n)^t = LZ \qquad (11.21)$$

The series $X_1, ..., X_n$ has the desired properties. In principle, this method is very easy to program. However, in the context of long memory one is mainly interested in generating rather long series. The covariance matrix is then very large. The Cholesky decomposition of Σ_n may therefore need a large amount of computer memory which might not be available. Also numerical problems might occur. It is therefore worthwhile to consider methods that avoid such large matrices. We briefly discuss some of these methods in the following sections.

11.3.2 Simulation of fractional Gaussian noise

In chapter 2, we saw that fractional Brownian noise and hence fractional Gaussian noise can be written as a stochastic integral with respect to Brownian motion. This suggests simulating fractional Gaussian noise by approximating this integral by a sum. As Brownian motion has independent increments, the only random numbers one needs to generate are iid standard normal variables. More details and several versions of this approximation are given in Mandelbrot (1971b) and Mandelbrot and Wallis (1969a).

11.3.3 A method based on the fast Fourier transform

Davies and Harte (1987) propose the following method for simulating a stationary Gaussian time series of length n with autocovariances $\gamma(0), \gamma(1), ..., \gamma(n-1)$:

1. Define
$$\lambda_k = \frac{2\pi(k-1)}{2n-2} \tag{11.22}$$
   for $k = 1, ..., 2n-2$, and the finite Fourier transform g_k of the sequence $\gamma(0), \gamma(1), ..., \gamma(n-2), \gamma(n-1), \gamma(n-2), ..., \gamma(1)$,
$$g_k = \sum_{j=1}^{n-1} \gamma(j-1)e^{i(j-1)\lambda_k} + \sum_{j=n}^{2n-2} \gamma(2n-j-1)e^{i(j-1)\lambda_k} \tag{11.23}$$
   for $k = 1, ..., 2n-2$.

2. Check that $g_k > 0$ for all $k = 1, ..., 2n-2$.

3. Simulate two independent series of zero mean normal random variables, say $U_1, U_2, ..., U_n$ and $V_2, ..., V_{n-1}$, such that
$$\text{var}(U_1) = \text{var}(U_n) = 2 \tag{11.24}$$
   and, for $k \neq 1, n$,
$$\text{var}(U_k) = \text{var}(V_k) = 1. \tag{11.25}$$
   Define $V_1 = V_n = 0$ and complex random variables Z_k by
$$Z_k = U_k + iV_k, \ k = 1, ..., n \tag{11.26}$$

and

$$Z_k = U_{2n-k} - iV_{2n-k} \qquad (11.27)$$

for $k = n+1, ..., 2n-2$.

4. For $t = 1, ..., n$, define

$$X_t = \frac{1}{2\sqrt{n-1}} \sum_{k=1}^{2n-2} \sqrt{g_k} e^{i(t-1)\lambda_k} Z_k. \qquad (11.28)$$

The series X_t has the desired distribution. The advantage of this method is that the fast Fourier transform can be used to calculate (11.23) and (11.28). This makes the method computationally fast.

11.3.4 Simulation by aggregation

A compationally efficient way of simulating long-memory processes follows from the results in 13.2. In a first step, one simulates a sufficient number of short-memory processes with parameters chosen randomly from a certain distribution. For instance, one may use AR(1) processes with coefficients α_j having a beta distribution (see Section 1.3.2). In a second step, the aggregated series (1.33) is obtained. In particular, this method is suitable for simulating extremely long time series. Because AR(1) processes can be calculated recursively, no numerical problems occur.

11.3.5 Simulation of fractional ARIMA processes

Fractional ARIMA(p, d, q) processes can be simulated in two steps, by using the representation (2.29).

1. Simulate a fractional ARIMA$(0,d,0)$ series $X_1^*, ..., X_n^*$ by one of the methods above.

2. Apply a program for the simulation of an ARMA process, using X_t^* as innovations.

The only difference from the simulation of an ordinary ARMA process is that instead of iid innovations the innovations X_t^* are used.

## CHAPTER 12

Programs and data sets

12.1 Splus programs

12.1.1 Simulation of fractional Gaussian noise

```
#######################
  #
  # Splus-functions for the simulation
  # of a series X(1),...,X(n) of
  # fractional Gaussian noise
  #
#######################

#######################
# ckFGN0
# covariances of a fractional Gaussian process
#——————————————————————-
# INPUT: n = length of time series
# H = self-similarity parameter
#
# OUTPUT: covariances up to lag n-1
#
#######################
ckFGN0 < −function(n,H)
    {
    k< −0:(n-1)
    H2< −2*H
    result< −(abs(k-1)**H2-2*abs(k)**H2+abs(k+1)**H2)/2
    drop(result)
    }


#######################
# gkFGN0
```

```
#——————————————————————
# A function to calculate gk=fft of V=(r(0),...,r(n-2),
# r(n-1), r(n-2),...,r(1), where r=the autocovariances
# of a fractional Gaussian process with variance 1
#
# INPUT: n = length of time series
# H = self-similarity parameter
#
# OUTPUT: gk = Fourier transform of V at
# Fourier frequencies
#
#####################################

gkFGN0 < −function(n, H)
    {
    gammak< −ckFGN0(n,H)
    ind < − c(0:(n - 2), (n - 1), (n - 2):1)
    gk < − gammak[ind+1]
    gk < − fft(c(gk), inverse = T)
    drop(gk)
    }


#################################
# simFGN0
#
# Simulation of a series X(1),...,X(n) of
# a fractional Gaussian process
#——————————————————————
# INPUT: n = length of time series
# H = self-similarity parameter
#
# OUTPUT: simulated series X(1),...,X(n)
#
#####################################

simFGN0 < −function(n,H)
    {
    z < − rnorm(2*n)
    zr < − z[c(1:n)]
    zi < − z[c((n+1):(2*n))]
    zic < − -zi
    zi[1] < − 0
```

```
zr[1] < - zr[1]*sqrt(2)
zi[n] < - 0
zr[n] < - zr[n]*sqrt(2)
zr < - c(zr[c(1:n)],zr[c((n-1):2)])
zi < - c(zi[c(1:n)],zic[c((n-1):2)])
z < - complex(real=zr,imaginary=zi)
cat("n=",n,"h=",H)
gksqrt < - Re(gkFGN0(n,H))
if(all(gksqrt>0))
  {
  gksqrt < - sqrt(gksqrt)
  z < - z*gksqrt
  z < - fft(z,inverse=T)
  z < - 0.5*(n-1)**(-0.5)*z
  z < - Re(z[c(1:n)])
  }else
  {
  gksqrt < - 0*gksqrt
  cat("Re(gk)-vector not positive")
  }
drop(z)
}
```

12.1.2 Simulation of fractional ARIMA(0, d, 0)

```
###################
#
# Splus-functions for the simulation
# of a series X(1),...,X(n) of
# a fractional ARIMA(0,d,0) process
# (d=H-1/2).
#
###################

###################
# ckARMA0
#
# covariances of a fractional ARIMA(0,d,0) process
#------------------------------------------------
# INPUT: n = length of time series
# H = self-similarity parameter
```

```
#
# OUTPUT: covariances up to lag n-1
#
###################

ckARMA0< -function(n,H)
   {
   result< -(0:(n-1))
   k< -1:(n-1)
   d< -H-0.5
   result[1]< -gamma(1-2*d)/gamma(1-d)**2

result[k+1]< -result[1]*(k**(2*H-2))*gamma(1-d)/gamma(d)
   k< -1:50
   result[k+1]< -result[1]*gamma(k+d)*
            gamma(1-d)/(gamma(k-d+1)*gamma(d))
   drop(result)
   }


###################
# gkARMA0
#─────────────────────────────
# A function to calculate gk=fft of V=(r(0),...,r(n-2),r(n-1),
# r(n-2),...,r(1)), where r=the autocovariances
# of a fractional ARIMA with innovation variance 0
#
# INPUT: n = length of time series
# H = self-similarity parameter
#
# OUTPUT: gk = Fourier transform of V at the
# Fourier frequencies
#
###################

gkARMA0< - function(n, H)
   {
   gammak< -ckARMA0(n,H)
   ind < - c(0:(n - 2), (n - 1), (n - 2):1)
   gk < - gammak[ind+1]
   gk < - fft(c(gk), inverse = T)
   drop(gk)
   }
```

```
##################
# simARMA0
#
# Simulation of a series X(1),...,X(n) of
# a fractional ARIMA(0,d,0) process (d=H-1/2)
#————————————————————
#
# INPUT: n = length of time series
# H = self-similarity parameter
#
# OUTPUT: simulated series X(1),...,X(n)
#
#################

simARMA0 < - function(n,H)
   {
   z < - rnorm(2*n)
   zr < - z[c(1:n)]
   zi < - z[c((n+1):(2*n))]
   zic < - -zi
   zi[1] < - 0
   zr[1] < - zr[1]*sqrt(2)
   zi[n] < - 0
   zr[n] < - zr[n]*sqrt(2)
   zr < - c(zr[c(1:n)],zr[c((n-1):2)])
   zi < - c(zi[c(1:n)],zic[c((n-1):2)])
   z < - complex(real=zr,imaginary=zi)

   cat("n=",n,"h=",H)
   gksqrt < - Re(gkARMA0(n,H))

   if(all(gksqrt>0))
      {
      gksqrt < - sqrt(gksqrt)
      z < - z*gksqrt
      z < - fft(z,inverse=T)
      z < - 0.5*(n-1)**(-0.5)*z
      z < - Re(z[c(1:n)])
      }else
      {
      gksqrt < - 0*gksqrt
      cat("Re(gk)-vector not positive")
      }
```

```
    drop(z)
    }
```

12.1.3 Whittle estimator for fractional Gaussian noise and fractional ARIMA (p, d, q)

```
####################
#
# Splus-functions and program for
# the calculation of Whittle's
# estimator and the goodness of
# fit statistic defined in Beran
# (1992). The models are
# fractional Gaussian noise or
# fractional ARIMA.
# The data series may be divided
# into subseries for which the
# parameters are fitted separately.
#
##################

#FFFFFFFFFFFFFFFF
# Functions
#FFFFFFFFFFFFFFFF

##################
# CetaFGN
#————————
# Covariance matrix of hat{eta}
# for fGn.
##################

CetaFGN< −function(eta)
    {
    M< −length(eta)

# size of steps in Riemann sum: 2*pi/m

    m< −10000
    mhalfm < − trunc((m-1)/2)

# size of delta for numerical calculation of derivative

    delta< −0.000000001
```

partial derivatives of log f (at each Fourier frequency)

```
lf< −matrix(1,ncol=M,nrow=mhalfm)
f0< −fspecFGN(eta,m)$fspec
for(j in (1:M))
{
  etaj< −eta
  etaj[j]< −etaj[j]+delta
  fj< −fspecFGN(etaj,m)$fspec
  lf[,j]< −log(fj/f0)/delta
}
```

Calculate D

```
Djl< −matrix(1,ncol=M,nrow=M)
for(j in (1:M))
  {for(l in (1:M))
    {Djl[j,l]< −2*2*pi/m*sum(lf[,j]*lf[,l])
    }
  }
```

Result

```
drop(matrix(4*pi*solve(Djl),ncol=M,nrow=M,byrow=T))
}
```

```
###################
# CetaARIMA
#———————
# Covariance matrix of hat{eta}
# for fractional ARIMA
###################

CetaARIMA< −function(eta,p,q)
  {
  M< −length(eta)
```

size of steps in Riemann sum: 2*pi/m

```
m< −10000    mhalfm < − trunc((m-1)/2)
```

size of delta for numerical calculation of derivative

```
delta< −0.000000001
```

partial derivatives of log f (at each Fourier frequency)

```
    lf< −matrix(1,ncol=M,nrow=mhalfm)
    f0< −fspecARIMA(eta,p,q,m)$fspec
    for(j in (1:M))
      {
      etaj< −eta
      etaj[j]< −etaj[j]+delta
      fj< −fspecARIMA(etaj,p,q,m)$fspec
      lf[,j]< −log(fj/f0)/delta
      }
# Calculate D

    Djl< −matrix(1,ncol=M,nrow=M)
    for(j in (1:M))
      {for(l in (1:M))
        {Djl[j,l]< −2*2*pi/m*sum(lf[,j]*lf[,l])
        }
      }
# Result

    drop(matrix(4*pi*solve(Djl),ncol=M,nrow=M,byrow=T))
    }


#################
# Qeta
#―――――――――――――――――
# Function for the calculation of A, B and
# Tn = A/B**2
# where A = 2pi/n sum 2*[I(lambda< −j)/f(lambda< −j)],
# B = 2pi/n sum 2*[I(lambda< −j)/f(lambda< −j)]**2 and
# the sum is taken over all Fourier frequencies
# lambda< −j = 2pi*j/n (j=1,...,(n-1)/2.
# f is the spectral density of fractional Gaussian
# noise or fractional ARIMA(p,d,q)
# with self-similarity parameter H=h.
# cov(X(t),X(t+k))=integral(exp(iuk)f(u)du)
#
# NOTE: yper[1] must be the periodogram I(lambda< −1) at
# ―- the frequency 2pi/n (i.e. not the frequency zero !).
#
# INPUT: h
#
```

```
# (n,nhalfm = trunc[(n-1)/2] and the
# nhalfm-dimensional vector yper must
# be defined.)
#
#
# OUTPUT: list(n=n,h=h,A=A,B=B,Tn=Tn,z=z,pval=pval,
# theta1=theta1,fspec=fspec)
#
# Tn is the goodness of fit test statistic
# Tn=A/B**2 defined in Beran (1992),
# z is the standardized test statistic,
# pval the corresponding p-value P(w>z).
# theta1 is the scale parameter such that
# f=theta1*fspec and integral(log[fspec])=0.
#
###################

Qeta < - function(eta)
   {
   cat("in function Qeta",fill=T)
   h < - eta[1]

# spectrum at Fourier frequencies

   if(imodel==1)
     {fspec < - fspecFGN(eta,n)
     theta1 < - fspec$theta1
     fspec < - fspec$fspec}
   else
     {fspec < - fspecARIMA(eta,p,q,n)
     theta1 < - fspec$theta1
     fspec < - fspec$fspec}

# Tn=A/B**2
   yf < - yper/fspec
   yfyf < - yf**2
   A < - 2*(2*pi/n)*sum(yfyf)
   B < - 2*(2*pi/n)*sum(yf)
   Tn < - A/(B**2)
   z < - sqrt(n)*(pi*Tn-1)/sqrt(2)
   pval < - 1-pnorm(z)
   theta1 < - B/(2*pi)
   fspec < - fspec
```

```
      Qresult< − list(n=n,h=h,
              eta=eta,A=A,B=B,Tn=Tn,z=z,pval=pval,
              theta1=theta1,fspec=fspec)
      drop(Qresult)
      }


#####################
# calculation of the spectral density f of
# normalized fractional Gaussian noise
# with self-similarity parameter H=h
# at the Fourier frequencies 2*pi*j/m (j=1,...,(m-1)).
#
# Remarks:
# ──── -
# 1. cov(X(t),X(t+k)) = integral[ exp(iuk)f(u)du ]
# 2. f=theta1*fspec and integral[log(fspec)]=0.
#
# INPUT: m = sample size
# h = self-similarity parameter
#
# OUTPUT: list(fspec=fspec,theta1=theta1)
#
#####################

fspecFGN < − function (eta,m)
      {
#──────parameters for the calculation of f──────
      h < − eta[1]
      nsum < − 200
      hh < − -2*h-1
      const < − 1/pi*sin(pi*h)*gamma(-hh)
      j < − 2*pi*c(0:nsum)

#────── x = 2*pi*(j-1)/m (j=1,2,...,(n-1)/2) ──────
#────── Fourier frequencies ──────────────-

      mhalfm < − trunc((m-1)/2)
      x < − 1:mhalfm
      x < − 2*pi/m*x

#────── calculation of f at Fourier frequencies ──────-

      fspec < − matrix(0,mhalfm)
```

```
for(i in seq(1:mhalfm))
{
lambda < − x[i]
fi < − matrix(lambda,(nsum+1))
fi < − abs(j+fi)**hh+abs(j-fi)**hh
fi[1] < − fi[1]/2
fi < − (1-cos(lambda))*const*fi
fspec[i] < − sum(fi)
}
```

#—- adjusted spectrum (such that int(log(fspec))=0 —

```
logfspec < − log(fspec)
fint < − 2/(m)*sum(logfspec)
theta1 < − exp(fint)
fspec < − fspec/theta1
drop(list(fspec=fspec,theta1=theta1))
}
```

```
####################
# calculation of the spectral density of
# fractional ARMA with standard normal innovations
# and self-similarity parameter H=h
# at the Fourier frequencies 2*pi*j/n (j=1,...,(n-1)).
# cov(X(t),X(t+k)) = (sigma/(2*pi))*integral(exp(iuk)g(u)du).
#
# Remarks:
# —— -
# 1. cov(X(t),X(t+k)) = integral[ exp(iuk)f(u)du ]
# 2. f=theta1*fspec and integral[log(fspec)]=0.
#
# INPUT: m = sample size
# h = theta[1] = self-similarity parameter
# phi = theta[2:(p+1)] = AR(p)-parameters
# psi = theta[(p+2):(p+q+1)] = MA(q)-parameters
#
# OUTPUT: list(fspec=fspec,theta1=theta1)
#
####################
```

```
fspecARIMA < − function (eta,p,q,m)
{
```

```
#———parameters for the calculation of f———

    cat("in fspecARIMA",fill=T)
    h < − eta[1]
    phi < − c()
    psi < − c()

#——— x = 2*pi*(j-1)/m (j=1,2,...,(n-1)/2) ———
#——— = Fourier frequencies ———————

    mhalfm < − trunc((m-1)/2)
    x < − 1:mhalfm
    x < − 2*pi/m*x

#——— calculation of f at Fourier frequencies ———-

    far < − (1:mhalfm)/(1:mhalfm)
    fma < − (1:mhalfm)/(1:mhalfm)
    if(p>0)
      {phi < − cbind(eta[2:(p+1)])
      cosar < − cos(cbind(x)%*%rbind(1:p))
      sinar < − sin(cbind(x)%*%rbind(1:p))
      Rar < − cosar%*%phi
      Iar < − sinar%*%phi
      far < − (1-Rar)**2 + Iar**2
      }
    cat("far calculated",fill=T)

    if(q>0)
      {psi < − cbind(eta[(p+2):(p+q+1)])
      cosar < − cos(cbind(x)%*%rbind(1:q))
      sinar < − sin(cbind(x)%*%rbind(1:q))
      Rar < − cosar%*%psi
      Iar < − sinar%*%psi
      fma < − (1+Rar)**2 + Iar**2
      }
    cat("fma calculated",fill=T)

    fspec < − fma/far*sqrt((1-cos(x))**2 + sin(x)**2)**(1-2*h)
    theta1 < − 1/(2*pi)
    cat("end of fspecARIMA",fill=T)
    drop(list(fspec=fspec,theta1=theta1))
    }
```

```
##################
# definition of the periodogram
##################

per < - function (z)
   {n< -length(z)
   (Mod(fft(z))**2/(2*pi*n)) [1:(n %/% 2 + 1)]
   }


##################
# definition of function to be minimized
##################

Qmin < - function (etatry)
   {
   result < - Qeta(etatry)$B
   cat("etatry=",etatry,"B=",result,sep=" ",fill=T)
   drop(result)
}


#MMMMMMMMMMMMM
# Main program
#MMMMMMMMMMMMM

# read data

      cat("in which file are the data ?")
      filedata < - readline()

      cat("total number of observations ?")
      nmax < - scan(n=1)

      cat("first and last observation to be considered (istart, iend)
?")
      startend < - c(scan(n=2)) # -- > we only look at
      istart < - startend[1] # observations
      iend < - startend[2] # istart,istart+1,...,end

      cat("into how many subseries do you divide the data ?")
      nloop < - scan(n=1)
      n < - trunc((iend-istart+1)/nloop)
      nhalfm < - trunc((n-1)/2)

# choose model
```

```
      cat("model: fr.G.noise (1) or fractional ARIMA(p,d,q) (2)
?")

      imodel < − scan(n=1)
      p < − 0
      q < − 0

      if(imodel==2)
        {
        cat("order of AR ?") #
        p < − scan(n=1)
        cat("order of MA ?") #
        q < − scan(n=1)
        }
  # initialize h

      cat("initial estimate of h=?")
      h < − scan(n=1)
      eta < − c(h)

  # initialize AR parameter

      if(p>0)
        {
        cat("initial estimates of AR parameters=?")
        eta[2:(p+1)] < − scan(n=p)
        }

  # initialize MA parameter

      if(q>0)
        {
        cat("initial estimates of MA parameters=?")
        eta[(p+2):(p+q+1)] < − scan(n=q)
        }
        M< −length(eta)

  # loop

      thetavector < − c()
      i0 < − istart
      for(iloop in (1:nloop))
        {
        h < − max(0.2,min(h,0.9)) # avoid extreme initial values
        eta[1] < − h
        i1 < − i0+n-1
```

```
    y < − c(scan(filedata,n=nmax))[i0:i1] # read only y[i0:i1]
# standardize data
    vary< −var(y)
    y< −(y-mean(y))/sqrt(var(y))
# periodogram of data
    yper < − per(y)[2:(nhalfm+1)]
# find estimate
    s < − 2*(1.-h)
    etatry < − eta
    result                        <
nlmin(Qmin,etatry,xc.tol=0.0000001,init.step=s)
    eta < − result$x
    theta1 < − Qeta(eta)$theta1
    theta < − c(theta1,eta)
    thetavector < − c(thetavector,theta)

# calculate goodness of fit statistic
    Qresult < − Qeta(eta)

# output
    M< −length(eta)
    if(imodel==1)
      {SD< −CetaFGN(eta)
      SD< −matrix(SD,ncol=M,nrow=M,byrow=T)/n
      }else
      {SD< −CetaARIMA(eta,p,q)
      SD< −matrix(SD,ncol=M,nrow=M,byrow=T)/n}
    Hlow< −eta[1]-1.96*sqrt(SD[1,1])
    Hup< −eta[1]+1.96*sqrt(SD[1,1])
    cat("theta=",theta,fill=T)
    cat("H=",eta[1],fill=T)
    cat("95%-C.I. for H: [",Hlow,",",Hup,"]",fill=T)

    etalow< −c()
    etaup< −c()
    for(i in (1:length(eta)))
      {
      etalow< −c(etalow,eta[i]-1.96*sqrt(SD[i,i]))
```

```
    etaup< −c(etaup,eta[i]+1.96*sqrt(SD[i,i]))
    }
cat("95%-C.I.:",fill=T)
print(cbind(etalow,etaup),fill=T)
cat("periodogram is in yper",fill=T)
fest < − Qresult$theta1*Qresult$fspec
cat("spectral density is in fest",fill=T)

# next subseries

    i0 < − i0+n
    }
```

12.1.4 Approximate MLE for FEXP models

```
####################
# Estimation of the parameters of a
# polynomial FEXP model as defined
# in Beran (1993).
# The order of the polynomial is
# increased until no significant
# improvement is achieved.
# The data series may be divided
# into subseries for which the
# parameters are fitted separatly.
#
####################

#FFFFFFFFFFFFFFFFFF
# Functions
#FFFFFFFFFFFFFFFFFFF

####################
# definition of periodogram
####################

per < − function (z)
    {n< −length(z)
    (Mod(fft(z))**2/(2*pi*n)) [1:(n %/% 2 + 1)]
    }

####################
# functions for plotting spectrum
```

```
###################

llplot < − function(yper,fspec)
  {plot(fglim,yper,log="xy")
  lines(fglim,fspec,col=4)
  }

lxplot < − function(yper,fspec)
    {plot(fglim,yper,log="y")
    lines(fglim,fspec,col=4)
    }


#MMMMMMMMMMMMMMMMM
# MAIN PROGRAM
#MMMMMMMMMMMMMMMMM

# read data

cat("in which file are the data ?")
filedata < − readline()

cat("total number of observations ?")
nmax < − scan(n=1)

cat("first and last observation to be considered (istart, iend) ?")
startend < − c(scan(n=2)) # −− > estimate for x[istart] to
x[iend]
istart < − startend[1]
iend < − startend[2]

cat("into how many subseries do you divide the data ?")
nloop < − scan(n=1)
n < − trunc((iend-istart+1)/nloop)

nhalfm < − trunc((n-1)/2)

cat("maximal order of polynomial ?")
p < − scan(n=1)

cat("P-value for entering new polynomial terms ?",fill=T)
pvalmax < − scan(n=1)

# Fourier frequencies

fglim< −(1:nhalfm)*2*pi/n
```

```
# loop (initialize)

thetamatrix < - matrix(0,ncol=p+2,nrow=nloop)
Hvector < - c()
fspecmatrix < - matrix(0,ncol=nhalfm,nrow=nloop)
ypermatrix < - matrix(0,ncol=nhalfm,nrow=nloop)
zero < - (1:(p+2))*0
i0 < - istart

# loop (start)

for(iloop in (1:nloop))
    {
    i1 < - i0+n-1
    y < - c(scan(filedata,n=nmax))[i0:i1]

# periodogram

    yper < - per(y)[2:(nhalfm+1)]

# long-memory component

    xlong< -log(sqrt((1-cos(fglim))**2 + sin(fglim)**2))

# fit with no polynomial

    xglim< -xlong

# generalized regression

    result<
-glim(xglim,yper,link="log",error="gamma",scale=0.5,resid="d")
    glim.print(result)

# estimate of theta

    theta< -result$coef
    eta< -theta[c(-1)]

# estimate of SD of theta

    SD< -result$var
    SD< -sqrt(SD[col(SD)==row(SD)])
    SD< -SD[c(-1)]

# estimated P-values

    pval< -2*(1-pnorm(abs(eta/SD)))
```

```
# estimate with no polynomials
   theta0< −theta
   xglim0< −xglim

# loop for choosing polynomial

   if(p>0){
     stopyesno< −"no"
     for(j in (1:p))
       {
       if(stopyesno=="no")
         {

# x-variables for long-memory and polynomial component

       xglim< −c()
       for(i in (1:j)){xglim< −cbind(xglim,fglim**i)}
       xglim< −cbind(xlong,xglim)

# generalized regression


result< −glim(xglim,yper,link="log",error="gamma",scale=0.5,
             resid="d")
       glim.print(result)

# estimate of theta

       theta< −result$coef
       eta< −theta[c(-1)]

# estimate of SD of theta

       SD< −result$var
       SD< −sqrt(SD[col(SD)==row(SD)])
       SD< −SD[c(-1)]

# estimated P-values

       pval< −2*(1-pnorm(abs(eta/SD)))

# condition for stopping

       if(max(pval)>pvalmax){stopyesno< −"yes"}
       else{theta0< −theta
         xglim0< −xglim}
       }
```

loop for choosing polynomial (end)

```
    }
  }
```

length of parameter vector

jmodel< −length(theta0)

estimate of theta for iloop'th subseries

thetamatrix[iloop,1:jmodel]< −theta0

estimate of H for iloop'th subseries

Hvector< −c(Hvector,(1-theta0[2])/2)

estimated spectral density

fspec < − exp(cbind(1,xglim0)%*%theta0)

estimated spectral density for iloop'th subseries

fspecmatrix[iloop,]< −c(fspec)

periodogram for iloop'th subseries

ypermatrix[iloop,]< −yper

output

cat("H=",Hvector[iloop],fill=T)
# loop (end)

```
    i0 < − i0+n
  }
```

cat("estimates are in thetamatrix, Hvector, ypermatrix, fspec-matrix",fill=T)
cat("For a plot of spectrum type",fill=T)
cat("llplot(yper,fspec) for log-log-plot",fill=T)
cat("lxplot(yper,fspec) for log-x-plot",fill=T)

12.2 Data sets

12.2.1 Nile River minima

Yearly minimal water levels of the Nile River for the years 622-1281, measured at the Roda Gauge near Cairo (Tousson, 1925, p.

366-385). The data are listed in chronological sequence by row.

1157	1088	1169	1169	984	1322	1178	1103	1211	1292
1124	1171	1133	1227	1142	1216	1259	1299	1232	1117
1155	1232	1083	1020	1394	1196	1148	1083	1189	1133
1034	1157	1034	1097	1299	1157	1130	1155	1349	1232
1103	1103	1083	1027	1166	1148	1250	1155	1047	1054
1018	1189	1126	1250	1297	1178	1043	1103	1250	1272
1169	1004	1083	1164	1124	1027	995	1169	1270	1011
1247	1101	1004	1004	1065	1223	1184	1216	1180	1142
1277	1206	1076	1076	1189	1121	1178	1031	1076	1178
1209	1022	1220	1070	1126	1058	1216	1358	1184	1083
1097	1119	1097	1097	1153	1153	1151	1151	1151	1184
1097	1043	1043	1002	1152	1097	1034	1002	989	1092
1115	1115	1047	1040	1038	1085	1126	1058	1067	1115
1263	1124	1110	1097	1097	1157	1000	991	995	1013
1007	971	971	980	993	1043	1097	982	971	971
1065	1022	1029	989	1029	995	982	1090	980	971
957	989	966	989	1022	1074	1110	1110	1061	1151
1128	1074	1043	1034	1074	966	1027	1029	1034	1065
989	1034	1002	1128	1178	1097	1142	1466	1097	1137
1097	1259	1313	1173	1169	1173	1088	1191	1146	1160
1142	1128	1169	1162	1115	1164	1088	1079	1083	1043
1110	1092	1110	1047	1076	1110	1043	1103	1034	1074
1052	1011	1097	1092	1110	1115	1097	1196	1115	1162
1151	1142	1126	1108	1187	1191	1153	1254	1187	1196
1331	1412	1349	1290	1211	1232	1166	1124	1146	1079
1108	1097	1106	1072	1065	1128	1340	959	959	1137
1133	1137	1151	1117	1157	1157	1133	1110	1155	1189
1260	1189	1151	1097	1209	1130	1295	1308	1250	1205
1310	1250	1155	1101	1100	1103	1121	1121	1097	1106
1259	1261	1124	1196	1205	1205	1119	1088	1250	1094
1198	1121	1164	1211	1153	1146	1126	1288	1175	1171
1081	1133	1164	1155	1155	1155	1160	1094	1054	1067
1044	948	1099	1016	1065	1067	1072	1076	1081	1196
1196	1151	1088	1128	1151	1236	1216	1288	1297	1182
1306	1043	1184	1054	1169	1043	980	1072	1189	1151

1142	1193	1151	1097	1144	1097	1094	1153	1108	935
1081	1081	1097	1146	1250	1151	1043	1043	1043	1070
1124	1137	1146	1099	1054	1045	1070	1142	1074	1101
1220	1196	1097	1207	1119	1160	1151	1025	1097	1137
1007	1034	1043	1043	980	1079	1169	1250	1324	1209
1142	1061	1000	1088	1128	1142	1259	1142	1148	1088
1142	1119	1130	1088	1250	1137	1108	1110	1173	1173
1196	1189	1200	1351	1274	1227	1310	1148	1151	1151
1182	1182	1151	1133	1130	1151	1166	1070	1200	1074
1110	1292	1178	1128	1097	1304	1103	1259	1119	1119
1119	1081	1196	1085	1101	1103	1146	1211	1169	1144
1191	1189	1182	1243	1243	1227	1189	1191	1155	1209
1218	1211	1209	1164	1135	1121	1137	1254	1457	1299
1277	1277	1178	1270	1313	1333	1270	1245	1245	1211
1265	1346	1346	1290	1295	1286	1259	1254	1421	1268
1263	1335	1313	1265	1319	1351	1277	1317	1268	1263
1112	1207	1292	1205	1223	1205	1153	1182	1245	1205
1151	1079	1151	1081	1128	1209	1157	1277	1259	1209
1220	1184	1220	1193	1247	1252	1259	1299	1173	1182
1180	1180	1331	1207	1236	1151	1182	1142	1191	1259
1166	1196	1241	1252	1241	1252	1157	1126	1164	1088
1173	1252	1288	1301	1286	1223	1232	1184	1207	1250
1256	1211	1216	1209	1209	1207	1151	1097	1097	989
966	1047	1056	1110	1290	1151	1166	1196	1196	1110
1110	1119	1119	1074	1106	1128	1218	1098	1044	1056
1058	1098	1043	1038	1142	1142	1193	1103	989	936
1142	1142	1151	1151	1180	1259	1196	1142	1169	1196
1142	1128	1043	1097	1142	1205	1205	1164	1160	1196
1112	1169	1110	1178	1133	1153	1139	1155	1187	1196
1220	1166	1128	1101	1157	1175	1142	1187	1254	1198
1263	1283	1252	1160	1234	1234	1232	1306	1205	1054
1151	1108	1097							

12.2.2 VBR data

VBR data (Heeke 1991, Heyman et al. 1991): original data (without log transformation). The data are listed in chronological sequence by row.

170	169	157	139	136	149	131	100	92	94
108	125	119	94	90	101	135	121	95	85
81	103	133	154	170	173	178	171	178	193
194	195	173	180	153	145	198	207	193	155
131	118	120	145	144	120	121	117	103	96
89	80	87	81	89	101	113	118	122	113
119	103	85	69	73	69	72	58	43	41
51	70	73	83	76	64	79	88	91	101
94	94	109	110	119	130	122	120	102	84
79	83	96	79	66	68	77	78	68	57
59	58	58	58	70	84	75	70	70	68
68	64	60	56	52	54	61	74	77	77
78	72	66	61	56	56	74	96	101	107
106	122	118	104	106	103	107	103	88	82
78	62	62	65	59	53	54	52	52	49
41	41	42	39	41	51	60	54	46	50
70	115	158	161	172	123	111	134	135	153
132	112	136	146	159	138	112	92	94	115
127	132	125	126	160	218	268	253	216	172
161	163	164	215	217	196	170	176	160	154
185	205	196	186	174	223	278	281	252	227
247	279	289	250	220	239	267	291	289	296
271	236	216	215	268	267	280	265	273	222
172	146	169	199	213	209	169	137	142	132
133	171	177	175	189	229	218	170	183	224
251	198	161	188	216	185	151	115	97	98
96	121	132	154	167	174	153	144	130	116
109	119	116	126	108	94	78	70	68	78
83	81	85	92	68	55	56	57	63	64
68	83	79	79	74	69	58	58	50	45
49	68	79	89	100	103	90	85	62	61
60	51	59	56	55	56	69	82	88	79
60	60	66	66	55	47	68	68	63	55
50	62	81	71	61	49	57	58	53	59
50	40	38	38	44	52	56	66	59	60

57	64	57	49	49	46	38	41	45	32
42	47	45	41	44	60	118	161	186	191
215	242	226	215	248	225	212	198	218	265
284	296	288	266	241	165	171	148	145	129
138	153	147	121	113	107	125	125	125	121
121	122	133	144	137	104	74	68	78	68
58	65	58	68	81	65	62	73	64	61
54	60	59	76	81	78	81	72	72	57
68	62	50	49	63	62	61	76	99	95
119	103	99	109	120	118	102	77	75	62
54	51	46	49	56	70	78	91	110	114
101	101	97	113	128	136	159	209	203	169
156	133	114	98	85	82	109	107	87	74
71	63	66	81	97	103	107	97	89	77
87	71	58	62	79	91	96	77	76	85
83	97	128	130	118	120	136	136	110	88
71	68	79	81	75	66	59	68	78	78
75	98	87	77	68	62	58	61	80	95
98	125	115	99	93	97	109	114	137	133
131	127	113	110	109	97	101	83	82	73
82	92	89	79	83	108	109	145	185	188
210	237	255	292	334	331	342	368	372	342
375	328	272	253	232	278	312	349	388	389
370	324	231	215	207	221	222	194	173	171
157	141	131	127	121	128	107	94	110	184
267	307	302	250	255	266	285	285	222	166
173	198	195	153	160	166	134	119	104	94
95	114	122	158	123	98	83	79	64	69
88	90	119	158	197	225	241	270	298	274
215	183	165	175	216	227	264	231	246	277
274	236	187	174	163	136	99	85	81	108
112	113	99	107	130	146	164	153	159	172
182	174	172	179	179	186	156	140	129	139
126	112	167	115	127	141	133	140	99	102
103	89	91	76	70	61	69	64	83	89

98	89	92	81	69	59	57	69	78	65
68	66	65	61	66	68	93	118	145	183
195	230	251	281	290	283	301	303	278	254
240	217	216	193	178	169	182	187	189	180
163	142	118	96	81	76	78	76	68	74
90	107	98	99	92	96	96	115	116	118
117	120	107	103	102	114	100	70	62	53
58	65	73	70	73	77	68	58	54	60
68	65	58	51	40	52	58	68	82	112
104	104	102	103	84	76	80	94	97	106
90	93	95	110	126	128	128	132	127	120
108	101	110	91	83	91	73	62	73	82
95	89	87	91	81	92	157	178	168	159
138	121	98	87	103	137	146	115	100	87
81	98	98	92	64	64	62	50	60	51
52	66	76	69	82	92	106	97	98	91
62	78	88	91	94	90	93	100	95	94
89	82	74	70	63	64	56	54	49	52
78	71	62	89	80	60	72	72	66	62
54	72	95	134	158	186	188	163	155	136
84	77	92	103	126	145	173	193	205	223
217	203	187	153	171	184	150	137	100	76
82	132	175	193	209	211	190	165	128	78
68	100	148	178	192	193	202	198	176	169
133	119	92	80	79	115	165	197	189	161
134	93	84	85	99	102	121	149	167	185
179	174	187	179	170	133	107	89	64	63
98	104	104	91	78	66	65	63	74	80
97	121	126	114	103	114	134	122	106	99
102	101	101	97	85	76	63	83	133	144

12.2.3 Ethernet data

Ethernet data from a LAN at Bellcore, Morristown (Leland et al. 1993, Leland and Wilson 1991). The data are listed in chronological sequence by row.

4858	5020	562	726	466	516	832	470	600	4076
5986	670	726	3978	6190	450	762	742	446	580
644	446	644	696	502	568	640	7128	2748	692
664	830	3630	10290	5132	438	796	466	568	6978
5982	6246	618	6738	2684	5974	7876	8004	5650	5820
5256	134	2356	308	198	302	302	220	1258	5748
2374	170	426	588	308	1368	232	194	102	232
130	102	232	130	228	114	592	248	1380	308
198	309	308	470	438	568	628	470	600	568
466	676	568	5916	4000	568	466	808	502	632
628	664	1196	832	713	2056	7756	672	648	900
2041	7480	696	9248	790	694	644	644	9038	802
534	672	2466	7552	734	2426	7318	9156	8836	806
6510	2922	660	374	130	4462	4658	272	168	232
608	232	6320	3000	512	232	4684	4076	402	4688
7140	5642	8642	770	296	1518	244	8662	264	450
198	110	390	300	110	244	8958	422	1498	194
236	3138	5724	588	778	568	628	5888	7694	5204
630	628	406	2118	7578	1038	524	754	4780	4796
750	632	446	9108	636	612	466	9148	822	9010
8028	1522	8994	2772	1082	1860	434	1004	568	1620
10442	6590	894	568	6344	9532	4984	9316	10822	12380
11200	6258	7332	11488	7191	8544	1208	650	110	588
270	7862	1600	366	308	7764	1238	252	236	6202
2558	398	8722	102	462	266	232	166	232	194
8630	130	360	166	228	130	98	244	134	114
244	134	174	3514	4560	1200	352	200	244	102
296	356	422	394	130	102	360	258	102	296
258	232	178	267	312	110	244	198	146	308
134	110	272	130	166	292	194	296	102	232
322	166	296	194	102	232	130	232	102	5746
3559	102	362	764	138	244	134	528	338	134
110	244	198	110	556	198	244	523	1196	624
502	794	8994	482	588	8662	272	308	134	3146
5736	292	9282	8370	714	1786	7318	478	238	372

198	5572	3584	4684	4142	232	284	406	394	472
1432	296	194	102	232	7720	1056	320	652	442
1510	134	178	312	134	110	244	470	612	580
470	942	668	470	672	644	470	620	636	530
794	502	568	5916	3948	634	692	502	568	640
690	820	692	596	438	742	466	438	794	594
434	768	584	612	584	470	450	470	754	450
644	696	510	580	644	110	308	262	110	308
134	102	232	194	568	5616	3530	272	106	448
626	612	604	420	354	322	130	98	750	194
1364	292	130	114	304	134	178	248	134	110
244	430	534	1440	244	264	110	244	134	446
580	296	620	580	696	502	568	6090	3774	640
694	438	632	640	438	820	718	600	568	466
676	674	692	438	568	640	632	438	628	568
434	768	712	514	746	470	446	818	534	778
580	470	620	580	470	742	470	418	134	5616
3568	130	168	232	194	166	480	782	484	178
248	134	110	248	198	174	308	198	110	244
352	706	166	130	1498	102	540	356	422	438
374	166	360	284	434	260	1396	248	2294	4738
2442	238	710	470	672	644	534	936	832	632
510	744	545	498	742	466	794	566	568	704
502	568	692	758	696	748	760	494	794	594
434	818	470	612	584	470	620	580	470	608
644	534	446	6030	3970	620	726	632	514	708
470	512	488	258	102	232	130	296	264	368
466	102	296	194	166	530	194	102	356	194
166	578	434	1432	228	130	114	248	231	244
110	198	520	174	308	6008	3672	1530	308	110
308	194	766	982	466	438	742	466	568	600
568	466	566	564	705	450	810	1290	506	584
644	648	608	580	470	620	580	534	600	636
530	568	676	568	594	1036	1152	6920	6396	7670
724	695	940	470	466	758	858	4306	450	998

308	264	308	490	552	244	134	174	534	372
176	300	4490	3588	620	130	1192	5682	2478	130
263	664	584	102	1638	832	178	727	733	252
342	134	174	396	300	5708	3798	460	1396	227
360	130	328	296	782	664	808	8244	1668	510
1399	594	676	564	1164	1180	858	1108	612	584
598	514	648	644	446	572	632	824	566	806
466	664	568	530	566	806	466	5892	4066	466
696	787	632	756	438	730	530	826	918	466
750	708	662	450	904	434	248	110	244	3170
5602	402	662	1628	7374	356	102	232	8658	102
390	194	4740	4246	288	398	3490	7080	292	210
64	64	128	128	5612	5514	3402	0	884	226
1330	128	638	0	346	970	1154	400	510	336
400	626	464	336	510	402	464	528	720	802
858	780	400	400	498	648	648	810	502	572
400	464	534	336	574	528	562	4696	4826	400
702	530	498	336	514	510	626	528	668	528
400	826	464	552	0	0	0	0	222	428
0	64	400	106	64	64	64	0	0	0
128	64	222	338	458	296	430	1340	64	0
0	64	64	64	192	5579	3336	0	362	162
1266	64	162	0	158	500	848	576	426	574
584	336	498	336	336	510	336	336	336	464
400	601	592	738	766	494	1016	668	732	602
248	498	510	162	336	336	635	400	498	336
400	400	400	5960	3672	336	628	400	336	510
336	652	814	400	400	510	0	0	192	253
256	64	192	64	64	162	174	0	64	64
406	478	240	64	64	0	64	64	64	424
162	1266	128	64	64	5450	3336	0	600	479
244	64	296	640	1420	189	0	66	0	336
336	498	464	400	336	510	336	562	400	400
510	400	464	810	806	668	584	336	1160	878
832	814	700	674	822	640	400	2755	4826	2580

464	692	428	336	336	510	588	588	498	410
478	400	646	604	2556	742	778	640	174	406
390	2244	74	158	402	628	414	0	248	440
388	665	202	548	414	240	0	0	154	4728
4426	1266	138	130	142	74	0	162	316	218
192	64	522	0	1266	189	64	64	0	336
400	464	562	336	336	510	400	562	336	336
510	924	474	464	138	324	128	256	446	0
142	5514	3584	64	0	130	138	142	0	236
128	154	226	90	0	0	0	0	0	0
0	0	248	0	0	0	0	0	256	315
166	0	206	0	0	64	352	364	64	142
0	64	128	8528	2338	5516	1154	7654	1004	0
0	0	213	290	64	252	0	154	128	64
0	192	192	304	129	206	0	0	0	0
0	0	128	0	0	0	64	248	64	0
0	64	0	64	0	0	0	0	64	64
0	5450	3400	0	0	66	0	0	64	226
0	316	712	336	426	574	400	336	562	336
336	510	400	336	498	336	336	510	400	498
336	400	400	510	336	336	498	336	336	510
336	336	498	336	410	510	336	0	0	5450
3336	64	194	0	0	0	498	0	154	162
227	64	64	0	192	166	0	142	64	296
162	1393	0	64	192	358	202	142	0	64
64	64	64	458	0	1266	74	142	64	0
0	400	562	336	410	510	336	5786	3834	592
594	640	652	400	498	498	336	588	510	426
336	498	400	464	822	336	336	498	336	400
411	510	336	498	336	400	336	510	336	336
498	461	336	336	584	510	336	498	336	238
0	0	0	0	81	5450	3336	226	304	0
0	64	64	162	0	252	0	90	0	0
218	304	1266	0	64	0	0	74	142	0
64	128	64	154	368	1266	0	0	0	74

0	336	400	562	336	461	336	336	830	656
738	542	400	5786	3672	510	400	402	562	336
336	746	400	652	498	426	574	464	464	464
754	592	585	336	550	638	400	498	400	400
574	400	400	648	464	336	510	400	562	174
0	64	0	0	0	0	0	64	400	64
0	5450	3400	66	0	64	0	128	456	64
446	466	1356	0	0	0	0	0	0	0
0	0	64	360	162	1266	0	64	67	124
0	400	400	498	336	336	336	510	336	498
336	336	400	5960	3736	562	336	466	336	510
336	498	562	426	498	920	490	400	562	336
336	574	336	562	400	400	336	510	336	562
336	336	336	510	400	464	746	400	336	174
0	64	192	190	64	0	336	0	64	128
64	5578	3400	0	130	64	248	0	438	64
154	290	154	64	1830	7034	64	458	0	1266
64	128	0	72	0	0	0	0	142	0
296	162	1266	64	0	0	0	0	336	562
336	336	336	510	336	464	498	455	336	5960
3672	400	722	400	336	510	336	2016	7508	746
336	490	574	8864	494	336	498	336	8864	494
400	574	336	8864	494	498	336	400	9001	732
562	464	464	528	6408	2966	494	128	71	3036
5492	222	64	1742	7244	332	0	3334	7590	6710
224	64	64	8592	350	226	64	698	4708	4368
1424	128	192	3036	5556	158	128	128	0	4682
3974	360	290	1266	128	128	0	64	464	535
498	238	698	400	336	510	400	464	498	336
336	574	419	626	336	5916	4288	736	974	708
174	562	400	747	336	426	660	336	426	336
336	633	336	498	400	336	464	574	336	562
336	400	400	336	574	400	562	174	0	64
0	72	270	0	128	128	0	0	584	0
0	64	64	5450	3464	194	64	64	0	64

0	522	316	1492	90	0	64	64	248	0
0	0	64	64	522	64	1266	0	64	64
64	226	336	510	336	336	498	336	336	336
510	400	400	689	336	336	668	336	400	6472
2792	6106	3800	0	8658	64	0	0	64	226
128	380	64	154	0	128	64	64	0	64
0	205	0	0	0	0	0	0	192	64
0	64	64	64	64	0	0	64	64	246
128	405	1992	1090	5514	2438	0	208	0	0
158	288	725	226	316	0	0	0	1582	7074
162	90	128	0	0	0	8528	0	64	0
130	64	64	1240	434	454	0	99	7654	938
0	64	64	226	0	0	8528	162	6072	2520
0	162	6200	2520	64	64	1751	11522	4274	128
162	6264	2584	0	0	3424	5628	64	0	162
8528	252	226	464	672	1854	7346	336	660	400
426	464	498	510	400	7990	7346	2954	336	336
336	672	4954	2804	162	308	64	64	64	0
0	0	0	64	71	150	0	0	0	240
248	5450	3644	434	430	300	66	0	0	3626
72	377	0	3334	5580	6630	2246	64	1156	1090
248	66	201	64	2244	64	3336	5516	64	128
1156	6606	1218	1156	1090	0	216	90	214	64
308	567	978	240	244	0	0	0	0	5450
3336	64	0	66	0	0	0	162	0	154
162	238	64	64	128	123	0	0	0	634
0	64	64	0	0	0	0	278	434	236
0	441	64	215	209	182	0	182	241	310
0	0	64	486	336	0	0	5774	3559	66
0	206	0	64	252	576	0	64	206	246
0	142	310	64	0	324	0	129	142	306
0	0	388	0	0	388	0	0	142	246
0	0	457	0	0	142	64	0	0	5450
3336	0	128	194	0	0	0	162	3522	5670
4360	5580	4360	2246	66	66	0	0	66	64

128	64	64	0	288	146	90	288	240	352
288	130	288	154	224	292	304	510	562	574
290	4360	4490	0	66	124	0	0	0	252
316	128	64	0	64	0	0	0	0	64
0	0	312	128	336	336	562	421	400	574
0	64	0	0	0	0	0	0	336	0
1090	5450	2558	0	66	0	0	0	0	226
342	296	162	1266	0	0	64	0	0	0
0	0	0	0	154	0	304	0	1394	64
0	0	0	162	0	0	0	128	132	0
0	0	0	0	64	5450	3336	0	66	0
0	64	376	702	284	0	67	64	530	3290
0	0	0	64	142	528	64	0	0	0
64	192	0	150	322	0	0	0	0	0
236	0	0	0	0	64	0	3270	4613	1394
134	288	0	196	0	570	252	822	0	0
64	350	0	0	64	0	0	0	0	64
0	64	0	0	0	0	0	0	0	0
0	0	0	0	0	0	64	0	0	5450
3336	66	0	0	0	0	252	316	0	0
0	0	0	226	400	424	0	510	336	400
336	380	336	442	336	174	290	400	336	0
510	336	336	162	510	336	336	162	400	464
336	5786	3672	336	336	402	174	336	336	498
588	252	174	498	422	162	64	336	0	400
174	162	174	310	336	336	336	336	336	174
498	174	498	336	248	510	336	336	400	64
64	64	187	5578	3464	64	130	400	0	0
0	252	252	64	0	0	64	64	64	64
0	360	162	0	1266	0	0	0	0	0
0	0	0	64	64	296	162	1266	64	0
128	166	4700	3998	64	5786	3672	336	628	238
454	562	400	588	762	640	336	336	510	336
336	400	562	336	336	510	336	336	498	336
336	510	464	690	336	400	510	336	336	562

336	400	510	336	1426	6014	2516	336	640	336
498	64	400	762	252	0	0	0	0	0
64	64	336	315	0	0	0	64	64	0
71	0	64	64	64	296	162	1330	64	64
0	192	64	0	0	0	0	0	296	5612
4602	0	130	0	0	162	498	852	336	498
336	400	510	400	336	498	336	336	510	336
336	498	336	336	510	336	562	400	400	510
336	336	498	336	336	336	510	336	400	753
408	5786	3910	402	626	336	762	174	626	794
652	400	528	400	238	0	0	0	162	530
128	64	226	238	64	0	248	0	0	192
64	64	0	64	135	154	368	1266	0	0
64	73	64	5450	3400	360	226	66	1404	72
119	226	380	380	409	336	336	510	400	626
336	336	400	336	510	336	498	336	336	574
336	464	336	498	336	336	336	400	0	510
336	336	162	174	336	5786	3834	336	466	574
336	336	498	498	900	510	401	498	3372	5828
622	464	574	1854	7346	158	0	64	0	4554
3974	600	304	430	8528	158	0	0	0	8528
0	558	0	3120	5556	222	0	5450	11864	0
288	64	64	8528	158	548	414	9794	222	0
0	8528	286	192	0	3332	5654	1424	0	64
0	8592	64	286	0	1808	7248	498	336	336
336	510	400	562	336	336	336	510	336	6012
3672	336	466	636	411	336	336	912	400	400
336	664	400	336	498	336	336	510	336	336
562	400	400	336	510	336	498	400	336	510
336	336	562	408	174	0	64	128	385	64
0	3398	5825	238	64	64	0	128	71	0
66	128	136	0	64	162	296	542	248	1420
828	270	128	0	8528	158	0	128	64	8528
616	0	1330	6136	2622	648	526	684	6308	2920
622	336	336	6408	2915	494	464	400	4890	4392

668	336	1744	7184	494	5786	6772	5956	320	66
64	2020	7074	320	426	426	162	7590	1096	146
154	304	9284	732	464	336	7926	1436	494	464
510	400	336	498	6408	2792	336	668	336	498
8864	336	336	574	400	460	336	3768	4762	1490
304	0	203	0	0	252	252	162	174	0
64	128	135	128	128	0	0	0	0	64
0	64	0	0	0	0	4350	296	162	0
0	64	1394	0	0	0	0	128	64	5592
3336	0	1584	2976	0	64	0	394	4687	0
142	4554	4086	64	214	452	64	64	0	64
0	128	128	0	0	0	0	0	0	0
0	0	458	64	1266	0	0	64	0	0
0	5578	3631	336	576	400	626	400	626	762
426	336	498	336	400	800	336	574	916	64
746	238	832	400	400	510	464	400	632	464
336	336	400	574	400	336	562	400	336	640
712	685	400	5786	3672	336	576	336	498	336
562	916	64	64	128	64	0	64	0	0
0	64	226	174	0	0	0	0	0	0
0	0	0	64	64	154	304	1266	0	0
0	64	0	128	236	5450	3336	130	0	296
226	0	1617	446	0	303	0	0	336	464
410	400	400	336	336	510	336	336	400	498
402	464	0	336	574	336	231	174	162	174
162	64	0	400	5786	3400	0	402	336	336
336	464	652	762	400	400	498	502	174	498
336	400	592	510	336	336	498	656	336	510
336	400	626	400	400	574	336	128	64	128
0	64	64	64	0	64	400	192	128	5450
3336	64	472	376	128	0	192	290	406	256
64	296	226	1512	0	364	248	334	192	64
64	190	128	64	64	522	0	1394	64	192
74	64	128	192	400	400	336	562	400	400
510	5786	3736	562	466	400	336	702	336	716

878	400	648	574	336	498	336	528	400	510
400	562	464	464	336	510	336	528	498	400
400	336	886	336	336	648	523	464	574	336
498	336	366	5450	3467	0	130	192	64	64
0	814	218	64	0	0	0	64	0	309
192	128	128	140	360	290	1266	64	0	64
64	256	0	0	64	0	64	128	128	154
368	3574	64	128	130	6540	3596	290	2180	6164
4760	10210	9186	4828	576	402	2434	402	628	1044
400	336	704	400	400	336	336	510	336	498
336	400	336	510	336	336	562	400	439	510
336	336	498	336	336	336	510	336	498	400
336	638	464	5786	3898	336	336	174	66	0
0	0	64	414	252	174	134	64	127	64
64	0	128	0	0	128	0	360	162	1330
0	0	64	64	134	0	0	64	64	107
296	226	1347	64	64	5514	3336	162	402	336
336	510	562	588	795	336	336	510	786	464
498	400	528	336	574	336	336	498	336	336
510	336	336	498	400	336	336	574	336	562
400	336	400	510	3854	4826	1588	336	402	574
336	498	498	588	426	510	0	191	0	0
64	128	64	64	474	174	0	0	64	64
0	64	64	64	0	0	0	218	368	1330
0	128	64	0	0	0	0	64	0	5450
3336	524	418	1330	64	64	380	380	336	498
336	336	336	574	528	886	563	4554	4516	2282
3136	2108	3200	498	464	400	414	574	336	690
400	400	528	574	400	336	498	336	336	510
482	5876	4074	400	564	466	510	636	576	652
824	554	746	482	720	628	400	174	64	64
0	64	128	242	226	174	0	64	0	0
0	64	64	0	256	65	0	0	296	162
1330	0	64	5450	3400	64	66	64	0	64
0	612	414	1266	0	0	64	0	400	562

400	400	574	414	400	562	336	464	746	576
498	564	336	510	400	400	498	464	465	510
336	336	498	526	336	5960	3672	400	402	562
336	336	638	750	588	336	336	612	336	336
498	336	336	336	510	400	0	0	0	0
128	64	0	400	0	64	377	64	0	0
0	64	0	64	64	64	522	5514	4679	64
130	64	0	0	162	252	634	162	0	1266
0	64	64	209	158	0	226	8928	494	510
336	8864	784	336	464	6310	2954	494	336	1854
7346	732	464	565	6536	2856	2012	7426	1854	10768
4610	940	336	336	498	8864	656	678	510	3372
5828	494	562	654	1918	7346	494	174	400	9572
7410	1274	1825	3326	304	574	162	1692	11490	336
558	238	0	7716	938	158	190	64	9770	5364
2372	4544	558	128	248	0	226	4660	318	64
0	0	4318	190	128	64	0	3036	1536	64
143	64	0	0	64	4318	1748	7442	146	64
206	162	3100	2802	0	0	142	128	64	4382
438	,0	64	5450	3400	1772	2928	256	0	0
0	4608	571	90	4446	550	162	1266	0	4440
190	0	128	0	0	1646	2962	428	226	1692
3026	364	336	1518	3390	64	4390	752	4492	400
162	4654	671	526	336	5960	3562	174	3566	1746
400	464	162	4906	968	716	336	302	336	4480
590	574	354	174	4480	364	336	498	260	498
64	8754	364	64	4544	590	3274	1698	3210	1872
336	162	238	4544	364	1518	2962	5734	3920	336
650	464	290	4654	703	750	264	4608	174	400
400	464	464	128	64	4318	190	0	4318	190
0	1518	2990	0	235	1518	2800	254	0	4382
648	64	354	4556	190	64	0	4318	190	5450
3464	0	322	192	4618	190	226	4698	344	0
1582	2990	0	0	3036	1661	0	0	0	4318
254	0	1710	2864	545	1646	3096	1836	3026	1584

4318	384	64	0	0	1518	2990	93	1518	2990
0	5450	3400	200	1698	2800	190	4318	352	342
0	1582	2800	486	162	1266	4318	190	0	1696
3374	0	64	1632	3293	0	0	4382	254	192
4446	318	64	128	1582	2990	336	0	4480	364
336	4318	352	5752	7721	4720	190	416	464	510
414	238	4480	526	494	174	1854	3388	4744	4136
4650	526	0	9098	4466	158	1854	3136	7780	2614
7280	780	464	10284	2128	4654	174	8811	336	336
336	8782	1854	3326	336	640	238	542	5472	6106
2516	400	530	400	1518	3200	2044	3376	854	4544
838	238	64	336	4480	526	400	400	174	4480
526	336	400	238	336	4608	492	0	1582	7186
312	4382	318	0	64	142	3036	5604	128	4318
376	128	0	5592	7954	4088	64	4318	254	336

12.2.4 NBS data

NBS weight measurements: deviation from 1 kg in micrograms (Pollak, Croarkin, and Hagwood 1993, Graf, Hampel, and Tacier 1984, Graf 1983). The data are listed in chronological sequence by row.

−19.48538	−19.58856	−19.35354	−19.41966	−19.51106
−19.46748	−19.42028	−19.474	−19.50776	−19.58184
−19.57892	−19.51508	−19.4894	−19.48844	−19.48178
−19.53168	−19.46304	−19.49595	−19.54338	−19.58635
−19.55626	−19.52787	−19.43625	−19.45378	−19.48934
−19.47218	−19.40403	−19.46627	−19.60931	−19.41321
−19.49616	−19.51869	−19.53728	−19.43859	−19.59881
−19.37827	−19.604	−19.53468	−19.47115	−19.61757
−19.44954	−19.44553	−19.46061	−19.41242	−19.42501
−19.374	−19.45432	−19.45692	−19.47184	−19.45627
−19.53346	−19.4675	−19.40454	−19.41526	−19.47652
−19.53466	−19.50872	−19.52944	−19.48906	−19.3921
−19.43192	−19.36784	−19.44396	−19.41736	−19.43518
−19.47676	−19.47532	−19.4663	−19.41172	−19.47366
−19.44744	−19.36256	−19.51832	−19.40878	−19.40388
−19.5131	−19.44582	−19.48936	−19.40668	−19.49002
−19.4771	−19.46216	−19.4966	−19.39358	−19.416
−19.54786	−19.44132	−19.56234	−19.43746	−19.51266
−19.35616	−19.40927	−19.39552	−19.3944	−19.41338
−19.42944	−19.40228	−19.47862	−19.4063	−19.53626
−19.45826	−19.5018	−19.48724	−19.451	−19.5628
−19.5234	−19.53438	−19.4892	−19.4438	−19.469
−19.4934	−19.4782	−19.4484	−19.53012	−19.4183
−19.49386	−19.41994	−19.44492	−19.42571	−19.48004
−19.37956	−19.47598	−19.4971	−19.42662	−19.53578
−19.5168	−19.52162	−19.45136	−19.53806	−19.4837
−19.45558	−19.51481	−19.50821	−19.52751	−19.45881
−19.50971	−19.4153	−19.59076	−19.48092	−19.49896
−19.43574	−19.4736	−19.44408	−19.44484	−19.45722
−19.45942	−19.36262	−19.48184	−19.5202	−19.42576
−19.4513	−19.47798	−19.41494	−19.47802	−19.4572
−19.5116	−19.4092	−19.5264	−19.50254	−19.49536
−19.4592	−19.4464	−19.3918	−19.5494	−19.55704
−19.48862	−19.38798	−19.58312	−19.469	−19.39354
−19.4856	−19.42142	−19.494	−19.3354	−19.446

−19.517	−19.473	−19.4778	−19.47672	−19.51816
−19.50724	−19.45304	−19.46672	−19.40318	−19.4835
−19.49608	−19.46312	−19.39336	−19.5266	−19.562
−19.4898	−19.545	−19.3722	−19.4404	−19.5264
−19.4176	−19.5116	−19.50038	−19.479	−19.4654
−19.5026	−19.5268	−19.4538	−19.48714	−19.49678
−19.50454	−19.50426	−19.5398	−19.50282	−19.47218
−19.46478	−19.5022	−19.47776	−19.45264	−19.47816
−19.46688	−19.43628	−19.52698	−19.41876	−19.51138
−19.4735	−19.47058	−19.4733	−19.4516	−19.5149
−19.5019	−19.5335	−19.5031	−19.4318	−19.5292
−19.5	−19.4751	−19.47329	−19.45363	−19.53239
−19.61721	−19.41739	−19.47822	−19.39572	−19.4928
−19.47579	−19.46452	−19.5128	−19.45849	−19.41639
−19.52435	−19.46989	−19.4668	−19.43825	−19.56234
−19.44987	−19.45597	−19.42954	−19.49172	−19.48178
−19.50401	−19.48206	−19.48174	−19.46688	−19.4653
−19.44681	−19.43597	−19.4793	−19.45009	−19.48532
−19.40451	−19.4398	−19.48486	−19.43898	−19.44598
−19.49492	−19.5402	−19.38595	−19.46289	−19.58291
−19.58791	−19.5579	−19.61791	−19.64791	−19.50789
−19.63291	−19.45465	−19.43605	−19.44945	−19.50779
−19.46674	−19.50868	−19.53614	−19.49073	

12.2.5 Northern hemisphere temperature data

Monthly temperature for the northern hemisphere for the years
1854-1989, from the data base held at the Climate Research Unit
of the University of East Anglia, Norwich, England (Jones and
Briffa 1992). The numbers consist of the temperature (degrees C)
difference from the monthly average over the period 1950-1979.
The data are listed in chronological sequence by row.

−0.88	−0.34	−0.07	−0.63	0.05	−0.54	0.54	0.02	−0.3	0.28
−0.53	0.51	−0.28	−0.79	−0.51	0.24	−0.13	−0.16	−0.13	−0.13
−0.41	−0.09	−0.78	−1.54	0.08	−0.51	−0.97	−0.62	−0.3	0.18
−0.13	−0.23	−0.23	−0.58	−0.92	−0.24	−0.52	−0.36	−0.27	−0.7
−0.76	−0.34	−0.18	0	−0.51	−0.64	−0.79	0.08	−0.33	−1.04
−0.65	−0.08	−0.16	−0.01	−0.11	−0.25	−0.21	0.11	−0.71	−0.55
−0.21	0.23	0.14	0.05	−0.01	−0.25	−0.21	0.01	−0.32	−0.03
−0.14	−0.51	−0.27	−0.72	−1.08	−0.45	−0.15	0.09	−0.08	−0.21
−0.18	−0.1	−1.08	−1.55	−1.62	−0.55	−0.04	−0.51	−0.53	−0.08
0.34	0.11	−0.13	−0.3	−0.22	−0.48	−1.33	−1.31	−0.49	−0.51
−0.44	−0.52	−0.42	−0.57	−0.32	−0.46	−1.17	−1.58	0.69	0.2
−0.15	−0.21	0	−0.3	−0.23	−0.28	−0.03	−0.44	−0.43	−0.34
−1.11	−0.52	−0.26	−0.46	−0.36	0.25	0.18	0.1	−0.46	−0.66
−0.67	−1.25	0.1	−0.96	−0.81	−0.44	−0.12	0.09	0.18	−0.11
0.27	−0.14	−0.04	−0.62	0.24	−0.24	−0.36	−0.18	−0.6	0.24
0.14	−0.21	−0.12	−0.62	−0.33	−0.27	−0.37	0.16	−0.55	−0.12
−0.75	−0.17	−0.2	−0.17	−0.14	−0.12	−0.11	−0.74	−0.69	−0.95
−0.07	−0.12	0.17	0.26	0.59	0.19	−0.33	−0.25	−0.68	−0.03
−0.29	0.69	−0.57	−0.26	0	0.17	0.17	0.33	0.12	−0.24
−0.39	−0.63	0.02	−0.68	−0.2	−0.13	0.12	0.28	0.12	−0.04
−0.12	−0.52	−0.06	−1.39	−0.81	−1.19	0.19	0.05	−0.25	−0.21
0.22	0.06	−0.53	−0.29	−0.78	−0.72	−0.58	−0.71	−0.31	−0.12
0.12	−0.03	0.12	0.11	0.03	−0.1	−0.49	−0.58	−0.19	−0.25
−0.37	−0.61	−0.39	−0.19	−0.08	−0.05	−0.25	−0.24	−0.36	−0.06
0.27	−0.28	−0.76	−0.42	−0.11	0	−0.13	−0.27	−0.08	−0.45
−0.58	−0.53	−0.87	−0.95	−0.92	−0.62	−0.18	0.09	−0.28	−0.15
−0.51	−0.58	−1.03	−0.83	−0.3	−0.23	−0.33	−0.25	−0.63	0
0.08	−0.18	−0.19	−0.47	−0.76	−1.18	−0.21	−0.22	−0.26	−0.45
−0.37	0.11	0.15	0.19	0.15	−0.09	0.21	0.06	−0.2	0.33
0.56	0.55	0.08	0.47	0.07	0.25	0.1	0.07	0.17	−0.37
−0.09	−0.02	−0.17	−0.33	0	−0.26	−0.2	−0.28	−0.35	−0.07
−0.72	−0.9	−0.18	−0.61	−0.33	−0.25	−0.17	−0.28	−0.24	−0.1
−0.32	−0.51	−0.63	−0.66	−0.54	−0.35	−0.37	−0.2	−0.01	−0.36
−0.2	−0.1	−0.34	−0.63	−0.7	−0.43	0.34	−0.06	0.01	−0.51
−0.52	−0.48	−0.32	−0.09	−0.17	−0.6	−0.59	−0.85	−0.84	−0.85

−0.49	−0.36	−0.41	−0.12	−0.17	−0.11	−0.42	−0.45	−0.58	−0.26
−0.27	−0.34	−0.67	−0.8	−0.5	−0.39	−0.42	−0.56	−0.61	−0.34
−0.91	−0.58	−0.87	−0.63	−0.28	−0.61	−0.62	−0.47	−0.25	−0.39
−0.34	−0.35	−0.43	−0.03	−0.65	−0.83	−0.47	−0.28	−0.18	−0.24
−0.14	−0.21	−0.22	−0.42	−0.57	−0.2	−0.82	−0.52	−0.26	−0.32
−0.25	−0.3	−0.13	−0.37	−0.22	−0.43	−0.19	−0.23	−0.55	−0.68
−0.48	−0.06	−0.2	−0.15	−0.17	−0.24	−0.23	−0.05	−0.17	−0.33
−0.48	−0.11	−0.04	0.13	−0.05	−0.03	−0.17	−0.18	−0.36	−0.41
−0.71	−0.35	−0.31	−0.32	−0.38	−0.21	−0.43	−0.32	−0.36	−0.43
−0.36	−0.36	−0.6	−0.44	−0.68	−0.74	−0.32	−0.4	−0.19	−0.18
−0.18	−0.19	−0.03	−0.26	−0.67	0.05	−0.53	−0.13	−0.6	−0.51
−0.37	−0.22	−0.38	−0.31	−0.16	−0.31	−0.78	−1.05	−1.71	−1.25
−0.36	−0.39	−0.68	−0.23	−0.06	−0.32	−0.23	−0.18	−0.29	−0.4
−0.7	−0.21	−0.15	−0.34	−0.32	−0.34	−0.26	−0.23	−0.34	−0.36
−0.51	−0.52	−0.93	−1.04	−0.62	−0.31	−0.31	−0.32	−0.26	−0.29
−0.08	−0.23	−0.21	−0.3	−0.25	−0.31	−0.61	−0.51	−0.04	0.06
−0.03	−0.01	−0.02	−0.05	−0.37	−0.14	−0.37	−0.33	−0.49	−0.01
0.1	−0.05	0.12	0.06	0.12	−0.05	−0.32	−0.57	0.23	−0.42
−1.11	−0.49	−0.33	−0.15	−0.18	−0.14	−0.14	−0.31	−0.15	−0.17
−0.01	−0.48	−0.45	−0.21	−0.15	−0.36	−0.08	−0.04	−0.06	0.07
0.36	−0.51	−0.44	−0.27	0.04	−0.13	−0.08	−0.06	−0.14	0.05
0.01	0.35	−0.11	0.08	−0.11	−0.04	0.18	0.16	−0.04	−0.05
0.04	0.07	−0.22	−0.21	−0.13	−0.48	−0.02	−0.05	−0.23	−0.46
−0.46	−0.37	−0.38	−0.22	−0.22	−0.35	−0.44	−0.53	0.09	0.25
−0.16	−0.46	−0.41	−0.52	−0.49	−0.49	−0.44	−0.5	−0.47	−0.58
−0.61	−0.44	−0.53	−0.58	−0.51	−0.51	−0.53	−0.39	−0.54	−0.38
−0.05	−0.24	−0.3	−0.86	−0.49	−0.62	−0.24	−0.33	−0.13	−0.19
−0.13	−0.2	0.12	−0.03	−0.19	−0.47	−0.13	0.01	−0.16	−0.13
−0.23	−0.11	−0.21	−0.23	−0.5	0.1	−0.44	−0.7	−0.36	−0.57
−0.65	−0.64	−0.37	−0.47	−0.34	−0.17	−0.6	−0.44	−0.32	−0.36
−0.74	−0.49	−0.47	−0.3	−0.4	−0.44	−0.32	−0.53	−0.55	−0.42
−0.63	−0.45	−0.62	−0.54	−0.45	−0.41	−0.39	−0.22	−0.21	−0.22
−0.03	−0.45	−0.16	−0.35	−0.29	−0.37	−0.41	−0.47	−0.32	−0.34
−0.42	−0.34	−0.55	−0.64	−0.54	−0.6	−0.56	−0.56	−0.48	−0.39
−0.34	−0.35	−0.24	−0.3	−0.19	−0.29	−0.3	−0.13	−0.46	−0.25

−0.22	−0.16	−0.47	−0.67	−0.66	−0.79	−0.6	−0.58	−0.45	−0.61
−0.41	−0.46	−0.63	−0.59	−0.62	−0.44	−0.44	−0.53	−0.13	0.03
0.19	−0.12	−0.26	−0.37	−0.2	−0.37	−0.35	−0.3	−0.26	0.06
−0.25	−0.27	−0.03	0.06	−0.25	0	−0.01	0	−0.03	−0.03
−0.02	−0.25	0.13	0	0.03	−0.17	−0.52	−0.24	−0.41	−0.37
−0.14	−0.25	−0.33	−0.3	−0.37	−0.87	−0.47	−0.87	−0.93	−0.55
−0.84	−0.35	−0.22	−0.21	−0.19	−0.5	−0.21	−1.26	−0.35	−0.4
−0.36	−0.74	−0.64	−0.54	−0.34	−0.4	−0.23	0	−0.3	−0.67
−0.29	−0.21	−0.5	−0.1	−0.5	−0.32	−0.33	−0.24	−0.16	−0.2
−0.71	−0.61	−0.03	−0.39	−0.03	−0.32	−0.17	−0.22	−0.23	−0.12
−0.23	−0.28	−0.59	−0.73	0.12	−0.23	−0.1	−0.11	−0.2	−0.07
−0.02	−0.22	−0.26	−0.19	−0.53	−0.24	−0.44	−0.45	−0.22	−0.32
−0.23	−0.16	−0.21	−0.2	−0.34	−0.25	−0.25	−0.24	−0.11	−0.67
−0.43	−0.62	−0.21	−0.29	−0.31	−0.32	−0.11	0.03	0.3	0.1
−0.35	−0.27	−0.23	−0.37	−0.15	−0.17	−0.18	−0.22	−0.16	−0.2
−0.17	−0.52	−0.27	−0.29	−0.15	−0.17	−0.17	−0.17	−0.12	−0.07
0.02	−0.18	0.02	0.24	0.51	0.25	0.24	−0.17	−0.12	−0.08
−0.06	−0.01	−0.02	0.06	0.14	−0.19	−0.09	0.02	−0.24	−0.21
−0.17	0	0.07	0.09	0.07	0.17	0.02	−0.47	0.23	0.04
−0.3	−0.29	−0.11	−0.2	−0.01	−0.12	−0.1	0.02	0.03	−0.09
−0.51	−0.97	−0.28	−0.36	−0.28	−0.24	−0.26	−0.12	−0.16	0.03
−0.01	−0.62	−0.3	−0.13	0.04	−0.12	−0.18	−0.17	0.08	0.16
−0.03	0.02	0.27	0.09	0.13	−0.33	0	−0.07	−0.11	0.15
0.1	0.16	0.15	0.23	0.03	0.05	0.45	−0.36	−0.28	0.02
−0.12	−0.01	−0.09	0.05	0.1	0.18	−0.23	−0.14	−0.4	−0.41
−0.36	−0.2	−0.27	−0.25	−0.02	−0.08	−0.15	−0.13	−0.24	−0.61
−0.25	0.09	−0.42	−0.24	−0.06	−0.07	−0.07	−0.14	−0.14	0.03
0.17	0.05	−0.41	0.52	−0.05	−0.29	−0.33	−0.12	−0.06	−0.06
−0.05	0.13	−0.42	−0.26	−0.25	−0.54	−0.15	−0.23	−0.06	0.02
0.14	0.05	0.02	−0.06	0.05	0.17	−0.12	0.08	−0.16	−0.06
0.01	0.13	0.17	0.22	0.3	0.38	0.18	−0.07	0.22	0.23
0.31	0.22	0.09	−0.03	0.08	0.2	0.3	0.36	0.21	−0.18
0.12	−0.08	−0.27	−0.09	−0.02	−0.03	0.09	0.1	0.02	−0.13
0.16	0.64	−0.34	0.01	−0.05	0.08	−0.11	−0.05	0.05	−0.09
0.16	0.02	0.05	0.19	0.01	0.33	0.03	0.05	−0.06	−0.08

0.05	−0.02	−0.17	0.18	−0.03	0.06	0.08	−0.11	0.21	0.1
0.05	0.06	0.01	−0.07	0.09	0.3	0.15	0.06	−0.22	0.22
−0.17	0.06	0.06	−0.15	0.08	0.11	0.08	0.38	0.27	0.44
0.6	0.4	0.26	0.07	0.21	0.15	0.08	0.08	0.31	0.36
0.09	−0.09	0.07	−0.09	0.18	0.06	−0.08	−0.06	−0.03	0.18
0.11	0.18	0.05	−0.33	0.49	0.28	0.15	0.26	−0.09	0.01
0.11	−0.04	0.12	−0.02	0.02	−0.4	−0.21	−0.03	0.21	0.12
−0.04	−0.05	0.06	0.05	0.01	0.27	0.19	−0.05	0.37	−0.12
−0.27	0.06	0.18	0.17	0.07	0.18	0.07	0.15	−0.04	−0.23
0.36	−0.14	−0.06	0.02	0.07	−0.08	−0.04	0.08	0	0.11
0.16	−0.18	−0.45	−0.29	0.07	−0.08	0.03	−0.06	−0.1	−0.14
0.03	−0.06	−0.32	0.01	−0.3	−0.51	−0.15	0.11	0.17	0.01
0.11	0.27	0.3	0.24	0.13	0.38	0.23	0.14	−0.14	0.19
0.14	0.2	0.18	0.13	0.16	0.03	−0.2	0.05	0.22	0.4
0.33	0.4	0.22	0.21	0.22	0.23	0.21	0.23	0.06	0.15
−0.35	−0.02	−0.1	−0.11	−0.08	0.01	−0.05	0.06	0.11	0.1
0.26	−0.17	0.43	−0.06	−0.43	−0.17	−0.01	0.01	0.03	0.12
−0.01	0.11	−0.14	−0.19	−0.01	−0.34	−0.32	−0.21	−0.26	−0.23
−0.18	−0.22	−0.28	−0.21	−0.28	−0.27	−0.12	−0.11	−0.09	−0.07
−0.05	0.07	0.08	0.16	0.14	0.04	0.17	0.35	0.59	0.33
0.27	0.11	0.1	0.02	0.05	0.12	0.07	0.07	0.1	0.22
0.32	0.19	0.27	0.17	0.03	0.15	0.08	0.07	0.16	−0.05
−0.14	0.02	0.1	0.54	−0.35	−0.13	0	0.17	0.1	0.1
0.12	0.05	−0.09	0.37	0.17	0.28	0.27	0.14	0.1	0.26
0.09	0.13	−0.02	−0.04	0.11	−0.05	0.32	0.43	0.24	0.21
0.12	−0.02	0.17	0.16	0.08	0.25	0.14	0.22	0.19	0.6
−0.09	0.04	0.01	−0.01	0.17	0.21	0.2	0.48	0.49	0.17
0.13	−0.09	−0.3	−0.26	−0.04	−0.05	−0.06	−0.21	−0.26	−0.25
−0.12	−0.3	−0.04	−0.41	−0.07	−0.29	−0.14	−0.19	−0.15	−0.25
−0.16	−0.02	−0.04	0	−0.09	0.06	0.05	−0.16	−0.04	0.07
0.13	0.15	0.12	0	−0.07	−0.2	−0.12	−0.24	0.15	0.05
0.16	−0.01	0.05	0.11	0.01	0.31	0.04	0.08	−0.26	−0.05
0.34	−0.02	0	−0.02	−0.08	−0.07	0.02	0.08	−0.06	−0.19
−0.3	−0.36	−0.05	0.1	0.02	−0.06	0.05	0.03	0.01	0.01
0.17	0.36	0.07	0.32	−0.03	0.09	0.01	0.02	−0.05	−0.03

−0.04	−0.15	0.03	−0.29	0.02	−0.26	−0.23	−0.18	−0.11	−0.18
−0.07	−0.19	−0.15	−0.13	0.06	−0.12	−0.61	−0.52	−0.14	−0.06
−0.13	−0.07	−0.05	−0.03	−0.27	−0.1	−0.18	−0.11	0.16	0.42
0.29	0.2	0.19	0.13	0.02	−0.01	−0.09	−0.05	−0.04	0.03
−0.44	−0.43	−0.11	−0.14	−0.21	−0.18	−0.04	−0.16	−0.16	−0.25
−0.17	−0.33	0.04	0.08	0.12	0.03	−0.03	−0.13	−0.12	−0.19
−0.1	−0.2	−0.31	−0.29	−0.02	−0.29	−0.48	−0.08	−0.29	−0.28
−0.31	−0.28	−0.22	−0.54	−0.3	−0.36	−0.24	0.07	0.26	0.23
0.17	0.16	0.01	−0.02	0.12	0.01	0.29	−0.05	0.18	0.15
0.16	0.07	−0.1	−0.18	−0.12	−0.16	−0.04	−0.02	0.21	−0.03
0.04	−0.3	0.16	−0.14	0.03	0.1	0.01	0.02	0.13	0.18
0.15	0.52	0.2	0.21	0.02	0.19	0.23	0.16	0.09	0.09
0.04	0.08	0.27	0.06	0.82	0.6	0.64	0.31	0.16	0.22
0.08	0.17	0.13	0.12	0.26	0.47	−0.11	0.19	−0.01	0.13
0.07	0.01	0.01	−0.07	0.02	0.01	−0.13	0.35	0.58	0.41
0.39	0.13	0.01	0.1	0.14	0.21	0.23	0.12	0.55	0.13
0.24	0.08	0.2	0	0.2	0.04	0.03	0.07	−0.07	−0.05
−0.21	−0.54	0.1	−0.29	0	−0.04	0.06	−0.06	−0.09	−0.04
−0.01	−0.02	−0.04	0.12	0.36	0.29	0.21	0.19	0.11	0.11
0	0.02	−0.02	0.04	−0.11	0.12	0.26	0.5	0.03	0.03
0.17	0.09	0.24	0.25	0.39	0.15	0.12	0.55	0.52	0.28
0.47	0.48	0.34	0.34	0.33	0.22	0.3	0.19	0.13	0.42
0.18	0.48	0.36	0.26	0.17	0.15	0.29	0.21	0.19	0.28
0.09	0.35								

Bibliography

The bibiliography contains representative references related to long memory. Not all references are referred to directly in the book, as in the book the focus is on certain selected topics. Because of the variety of applications, the list is far from comprehensive. Some selected references to related topics, such as model choice or infinite variance processes, are also given.

Abraham, B. and Ledolter, J. (1983) Statistical methods for forecasting. *Wiley, New York*.

Adelman, I. (1965) Long cycles - fact or artifact? *Am. Econ. Rev.*, **60**, 444-463.

Adenstedt R.K. (1974) On large sample estimation for the mean of a stationary random sequence. *Ann. Statist.*, **2**, 1095-1107.

Adler, R.J. (1981) The geometry of random fields. *Wiley, New York*.

Agiakloglou, C., Newbold, P., and Wohar, M. (1994) Lagrange multiplier tests for fractional difference. *J. Time Ser. Anal.*, **15, No. 3**, 253-262.

Agiakloglou, C., Newbold, P., and Wohar, M. (1993) Bias in an estimator of the fractional difference parameter. *J. Time Ser. Anal.*, **14, No. 3**, 235-246.

Akaike, H. (1960) On a limiting process which asymptotically produces f^{-2} spectral density. *Ann. Inst. Statist. Math.*, **12**, 7-11.

Akaike, H. (1969) Fitting autoregressive models for prediction. *Ann. Inst. Statist. Math.*, **21**, 243-247.

Akaike, H. (1973a) Information theory and an extension of the maximum likelihood principle. *2nd Int. Symp. on Information Theory, B.N. Petrov and F. Csaki (eds.), Akademiai Kiado, Budapest*, 267-281.

Akaike, H. (1973b) Maximum likelihood identification of Gaussian autoregressive moving average models. *Biometrika*, **60**, 255-265.

Akaike, H. (1978) A Bayesian analysis of the minimum AIC procedure. *Ann. Inst. Statist. Math.*, **30A**, 9-14.

Akaike, H. (1979) A Bayesian extension of the minimum AIC procedure of autoregressive model fitting. *Biometrika*, **66**, 237-242.

Akashi, S. (1992) A relation between Kolmogorov-Prokhorov's condition and Ohya's fractal dimensions. *IEEE Trans. Inf. Th.*, **38**, 1567-1583.

Akcasu, A.Z. (1961) Measurement of noise power spectra by Fourier analysis. *Ann. Statist.*, **2**, 1095-1107.

Akonom, I.A. and Gourieroux, C. (1987) A functional limit theory for fractional processes. *preprint*.

Albers, A. (1978) Teting the mean of a normal population under dependence *Ann. Statist.*, **6**, 261-269.

Aldous, I.D. (1991) The continuum random tree. *Ann. Probab.*, **19**, 1-28.

Allan, D.W. (1966) Statistics of atomic frequency standards. *Proc. IEEE*, **54**, 221-230.

An, H.Z. and Chen, Z.G. (1982) On convergence of LAD estimates in autoregression with infinite variance. *J. Multiv. Anal.*, **12**, 335-345.

Andel, J. (1986) Long memory time series models. *Kybernetika*, **22**, 105-123.

Andel, J. and Gomez, M. (1988) Two-dimensional long-memory models. *Kybernetika*, **24, No. 1**, 1-16.

Anderson, O.D. (1990) Small-sample autocorrelation structure for long-memory time series. *J. Oper. Res.*, **41**, 735-754.

Anis, A.A. and Lloyd, E.H. (1976) The expected values of the adjusted rescaled Hurst range of independent normal summands. *Biometrika*, **63**, 111-116.

Arbeiter, M. (1991) Random recursive construction of self-similar fractal measures. The noncompact case. *Probab. Th. Rel. Fields*, **88**, 497-520.

Arbeiter, M. (1992) Construction of random fractal measures by branching processes. *Stochastics*, **39**, 195- 212.

Arcones, M.A. and Yu, B. (1992) Empirical processes under long-range dependence. *preprint*.

Ash, R.B. and Gardner, M.F. (1975) Topics in stochastic processes. *Academic Press, New York*.

Astrauskas, J. (1983a) Limit theorems for sums of linearily generated random variables. *Litovsk. Mat. Sb.*, **23, No. 2**, 3-12.

Astrauskas, A. (1983b) Limit theorems for quadratic forms of linear processes (Russian). *Litovsk. Mat. Sb.*, **23, No. 4**, 3-11.

Astrauskas, J., Lévy, J.B., and Taqqu, M.S. (1992) The asymptotic dependence structure of the linear fractional Lévy motion. *Litovskii Matematicheskii Sbornik*, **31, No. 1**, 3-28.

Avram F. (1987) Asymptotics of sums with dependent indices and convergence to the Gaussian distribution. *preprint*.

Avram F. (1988a) On bilinear forms in Gaussian random variables and Toeplitz matrices. *Probab. Th. Rel. Fields*, **79**, 37-45.

Avram, F. (1988b) Generalized Szegö theorems and asymptotics of cumulants by graphical methods. *preprint*

Avram F. (1988c) On the central limit theorem for sums of Wick powers. *preprint*.

Avram F. and Brown L. (1989) A generalized Hölder inequality and a generalized Szegö theorem. *Proc. Am. Math. Soc.*, **107, No. 3**, 687-695.

Avram, F. and Fox, R. (1992) Central limit theorems for sums of Wick products of stationary sequences. *Trans. Am. Math. Soc.*, **330**, 651-663.

Avram, F. and Taqqu, M.S. (1986) Weak convergence of moving averages with infinite variance. *Dependence in Probability and Statistics, E.Eberlein and M.S.Taqqu (eds.), Birkhäuser, Boston*, pp. 399-415.

Avram, F. and Taqqu, M.S. (1987) Noncentral limit theorems and Appell polynomials. *Ann. Probab.*, **15**, 767-775.

Aydogan, K. and Booth, G.G. (1988) Are there long cycles in common stock returns ? *Southern Economic Journal*, 141-149.

Bailey, R.A. (1985) Restricted randomization versus blocking. *Int. Statist. Rev.*, **53**, 171-182.

Bailey, R.A. and Rowley, C.A. (1987) Valid randomization. *Proc. Roy. Soc. London A*, **410**, 105-124.

Bak, P., Tang, C., and Wiesenfeld, K. (1987) Self-organized critically: An explanation of $1/f$ noise. *Phys. Rev. Lett.*, **59**, 381-384.

Baldessari, B. (1987) Discussion of session on self-similar processes. *Proceedings of the 46th Session of ISI, Tokyo, Book 4*, p. 261.

Bandt, C. (1989) Self-similar sets. I. Topological Markov chains and mixed self-similar sets. *Math. Nachr.*, **142**, 107-123.

Barlow, M.T. and Bass, R.F. (1990) Local times for Brownian motion on the Sierpinski carpet. *Probab. Th. Rel. Fields*, **85**, 91-104.

Barlow, M.T. and Perkins, E.A. (1988) Brownian motion on the Sierpinski gasket. *Probab. Th. Rel. Fields*, **79**, 543-623.

Barnsley, M. (1988) Fractals everywhere. *Academic Press, New York.*

Barnsley, M.F. and Elton, J.H. (1988) A new class of Markov processes for image encoding. *Adv. Appl. Probab.*, **20**, 14-32.

Bartlett, M.S. (1948) Smoothing periodograms from time series with continuous spectra. *Nature*, **161**, 686-687.

Bartlett, M.S. (1950) Periodogram analysis and continuous spectra. *Biometrika*, **37**, 1-16.

Barton, R.J. and Poor, H.V. (1988) Signal detection in fractional Gaussian noise. *IEEE Trans. Inf. Th.*, **34**, 943-995.

Basseville, M. Detecting changes in signals and systems - a survey. *Automatica*, **24**, 309-326.

Batchelor, G.K. (1953) The theory of homogeneous turbulence. *Cambridge University Press, Cambridge.*

Batty, M. and Longley, P.A. (1986) The fractal simulation of urban structure. *Envir. Pl. A*, **18**, 1143-1179.

Beauregard, J. (1968) Débits mensuels et modules des fleuves francais. *Electricité de France, March 1968.*

Beck, U. (1987) Computer graphics. Pictures and programs for fractals, chaos, and self-similarity (German). *Birkhäuser, Basel.*

ben-Avraham, D. (1991) Diffusion in disordered media. *Chemom. Int. Lab. Syst.*, **10**, 117-122.

Beran, J. (1984) Maximum likelihood estimation for Gaussian processes with long-range dependence. *Abstracts, 16th European Meeting of Statisticians, Marburg, West Germany*, p. 71.

Beran, J. (1986) Estimation, testing and prediction for self-similar and related processes. *PhD Thesis, ETH Zürich.*

Beran, J. (1988) Statistical aspects of stationary processes with long-range dependence. *Mimeo Series No. 1743, Department of Statistics, University of North Carolina, Chapel Hill, NC.*

Beran, J. (1989a) A test of location for data with slowly decaying serial correlations. *Biometrika*, **76**, 261-269.

Beran, J. (1989b) Akaike's model choice criterion, long-range dependence and approximate maximum likelihood estimation. *Tech. Rep., No. 96, Department of Statistics, Texas A&M University.*

Beran J. (1991) M-estimators of location for data with slowly decaying serial correlations. *J. Am. Statist. Assoc.*, **86**, 704-708.

Beran, J. (1992a) A goodness of fit test for time series with long-range dependence. *J. Roy. Statist. Soc., Series B***54, No. 3**, 749-760.

Beran, J. (1992b) Statistical methods for data with long-range dependence. (with discussion) *Statist. Sci.*, **7, No.4**, 404-416.

Beran, J. (1992c) Rejoinder to discussion of "Statistical methods for data with long-range dependence." *Statist. Sci.*, **7, No.4**, 425-427.

Beran, J. (1992d) Recent developments in location estimation and regression for long-memory processes. *Directions in Time Series Analysis, Part II, D. Brillinger, P. Caines, J. Geweke, E. Parzen, M. Rosenblatt, and M.S. Taqqu (eds.), IMA Volumes in Mathematics and its Applications, Vol. 46, Springer, New York*, pp. 1-9.

Beran, J. (1992e) Modelling "Starwars" and other video data. *Royal Statistical Society Conference, Abstracts, Sheffield, UK.*

Beran, J. (1993a) Fitting long-memory models by generalized linear regression. *Biometrika, Vol. 80, No. 4*, 817-822.

Beran, J. (1993b) On a class of M-estimators for long-memory Gaussian models. *preprint.*

Beran, J. (1994) Maximum likelihood estimation of the differencing parameter for invertible short- and long-memory ARIMA models. *preprint.*

Beran, J. and Ghosh, S. (1990) Goodness of fit tests and long-range dependence. *Directions in Robust Statistics and Diagnostics, Part I, W. Stahel and S. Weisberg (eds.), IMA Volumes in Mathematics and its Applications, Vol. 33, Springer*, pp. 21-33.

Beran, J. and Ghosh, S. (1991) Slowly decaying correlations, testing normality, nuisance parameters. *J. Am. Statist. Assoc.*, **86**, 785-791.

Beran, J. and Künsch, H.R. (1985) Location estimators for processes with long-range dependence. *Research Report No. 40, Seminar für*

Statistik, ETH Zürich.

Beran, J., Sherman, R., Taqqu, M.S., and Willinger, W. (1992) Variable-Bit-Rate Video Traffic and Long-Range Dependence. *preprint, Bellcore, Morristown, NJ.*

Beran, J. and Terrin, N. (1994) Estimation of the long-memory parameter, based on a multivariate central limit theorem. *J. Time Ser. Anal., Vol. 15, No. 3,* 269-278.

Berenblut, I.I. and Webb, G.J. (1974) Experimental designs in the presence of autocorrelated errors. *Biometrika,* **61,** 427-437.

Berger, M.A. (1989) Images generated by orbits of 2-D Markov chains. *Chance,* **2, No. 2,** 18-28.

Berkson, J. (1938) Some difficulties of interpretation encountered in the application of the chi-square test. *J. Am. Statist. Assoc.,* **33,** 526-536.

Berliner, L.M. (1992) Statistics, probability and chaos. (with discussion) *Statist. Sci.,* **7,** 69-90.

Berliner, L.M. (1992) Reply to comments on "Statistics, probability and chaos." *Statist. Sci.,* **7,** 118-122.

Berman, S.M. (1979) High level sojourns for strongly dependent Gaussian processes. *Z. Wahrsch. verw. Geb.,* **50,** 223-239.

Berman, S.M. (1984) Sojourns of vector Gaussian processes inside and outside spheres. *Z. Wahrsch. verw. Geb.,* **66,** 529-542.

Berry, M.V. (1979) Diffractals. *J. Phys. A,* **12,** 781-797.

Bhansali, R.J. (1984) Order determination for processes with infinite variance. *Robust and Nonlinear Time Series Analysis. J. Franke, W.Härdle, R.D. Martin (eds.), Lecture Notes in Statistics, Vol. 26, Springer, New York,* pp. 17-25.

Bhansali, R.J. (1988) Consistent order determination for processes with infinite variance. *J. Roy. Statist. Soc., Series B,* **50,** 46-60.

Bhattacharya, R.N., Gupta, V.K., and Waymire, E. (1983) The Hurst effect under trends. *J. Appl. Probab.,* **20,** 649-662.

Billingsley, P. (1968) Convergence of probability measures. *Wiley, New York.*

Bingham, N.H., Goldie, C.M. and Teugels, J.L. (1987) Regular variation. *Cambridge University Press, Cambridge.*

Bleher, P.M. (1981) Inversion of Toeplitz matrices. *Trans. Moscow Math. Soc.,* **2,** 201-229.

Bloomfield, P. (1973) An exponential model for the spectrum of a scalar time series. *Biometrika,* **60,** 217-226.

Bloomfield, P. (1992) Trends in global temperature. *Climatic Change,* **21,** 1-16.

Bloomfield, P. and Nychka, D. (1992) Climate spectra and detecting climate change. *Climatic Change,* **21,** 275-287.

Bollerslev, T. and Engle, R.F. (1993) Common persistence in conditional variances. *Econometrica,* **61,** 167-186.

Boes, D.C., Davis, R.A., and Gupta, S.N. (1989) Parameter estimates in low order fractionally differenced ARIMA processes. *Stochast. Hydrol. Hydraul.*, *3*, 97-110.

Boes, D.C. and Salas-LaCruz, J.D. (1973) On the expected range of partial sums of exchangeable random variables. *J. Appl. Probab.*, **10**, 671-677.

Boes, D.C. and Salas-LaCruz, J.D. (1978) Nonstationarity of the mean and the Hurst phenomenon. *Water Resources Res.*, **14**, 135-143.

Booth, G., Kaen, F., and Koveos, P. (1982) R/S analysis of foreign exchange rates under two international monetary regimes. *J. Monet. Econ.*, **10**, 407-415.

Boulos, N. (1951) Probability in the hydrology of the Nile. *Cairo, Government Press.*

Box, G.E.P., Hunter, W.G. and Hunter, J.S. (1978) Statistics for experimenters. *Wiley, New York.*

Box, G.E.P. and Jenkins, G.M. (1970) Time series analysis: forecasting and control. *Holden Day, San Francisco.*

Box, G.E.P. and Ljung, G.M. (1978) Lack of fit in time series models. *Biometrika*, **65**, 297-303.

Box, G.E.P. and Pierce, D.A. (1970) Distribution of residual autocorrelations in autoregressive moving average time series models. *J. Am. Statist. Assoc.*, **65**, 1509-1526.

Bramson, M. and Griffeath, D. (1979) Renormalizing the 3-dimensional voter model. *Ann. Probab.*, **7**, 418-432.

Bretschneider, C.L. (1959) Wave variability and wave spectra for wind-generated gravity waves. *U.S. Army Corps of Engineers, Beach Erosion Board, Techn. Memo. No. 118.*

Breuer, P. and Major P. (1983) Central limit theorems for non-linear functionals of Gaussian fields. *J. Multivar. Anal.*, **13**, 425-441.

Brockwell, P.J. and Davis, R.A. (1987) Time series: Theory and Methods. (first edition). *Springer, New York.*

Broomhead, D.S., Huke, J.P., and Muldoon, M.R. (1992) Linear filters and non-linear systems. *J. Roy. Statist. Soc., Series B*, **54**, 373-382.

Brox, Th. (1986) A one-dimensional diffusion process in a Wiener medium. *Ann. Probab.*, **14**, 1206-1218.

Burdzy, K. (1989) Cut points on Brownian paths. *Ann. Probab.*, **17**, 1012-1036.

Burdzy, K. and Lawler, G.F. (1990) Nonintersection exponents for Brownian paths. II. Estimates and applications to a random fractal. *Ann. Probab.* , **18**, 981-1009.

Burrough, P.A. (1981) Fractal dimensions of landscapes and other environmental data. *Nature*, **294**, 240-242.

Burton, R. and Waymire, E. (1983) Limit theorems for point random fields. *preprint.*

Cabrera, J. and Cook, D. (1992) Projection pursuit indices based on fractal dimension. *Proc. Comp. Sci. and Statist., 9th Annual Symp.*

on the Interface, UCLA, Vol. 24, pp. 474-477.

Cahalan, R.F. and Joseph, J.H. (1989) Fractal statistics of cloud fields. Monthly Weather Rev., **117**, 261-272.

Cambanis, S. (1983) Symmetric stable variables and processes. Contributions to Statistics, Essays in Honor of Norman L. Johnson, P.K. Sen (editor), North Holland, pp. 63-79.

Cambanis, S., Hardin, Jr., and Weron, A. (1988) Innovations and Wold decomposition of stable sequences. Probab. Th. Rel. Fields, **79**, 1-27.

Cambanis, S. and Maejima, M. (1989) Two classes of self-similar stable processes with stationary increments. Stoch. Proc. Appl., **32**, 305-329.

Cambanis, S., Maejima, M., and Samorodnitsky, G. (1992) Characterization of linear and harmonizable fractional stable motions. Stoch. Proc. Appl., **42**, 91- 110.

Cameron, M.A. and Turner, T.R. (1987) Fitting models to spectra using regression packages. Appl. Statist., **36, No.1**, 47-57.

Campbell, J.Y. and Mankiw, N.G. (1987) Are output fluctuations transitory ? Quaterly Journal of Economics, **102**, 857-880.

Carlin, J.B. (1987) Seasonal analysis of economic time series. PhD thesis, Dept. of Statistics, Harvard University

Carlin, J.B. and Dempster, P. (1989) Sensitivity analysis of seasonal adjustments: Empirical case studies. J. Am. Statist. Assoc., **84**, 6-20.

Carlin, J.B., Dempster, P., and Jonas, A.B. (1985) On methods and models for Bayesian time series analysis. J. Econometrics, **30**, 67-90.

Carlin, J.B. and Haslett, J. (1982) The probability distribution of wind power from a dispersed array of wind turbine generators. J. Appl. Meteorol., **21**, 303-313.

Carr, J.R. and Benzer, W.B. (1991) On the practice of estimating fractal dimension. Math. Geol., **23**, 945-958.

Carter, P.H., Cawley, R., and Mauldin, R.D. (1988) Mathematics of dimension measurement for graphs of functions. Fractal aspects of materials. Weitz, D.A., Sander, L.M. and Mandelbrot, B.B. (eds.), Material Res. Soc., Pittsburgh, pp. 183-186.

Cassandro, M. and Jona-Lasinio, G. (1978) Critical behavior and probability theory. Adv. Physics, **27**, 913-941.

Chambers, E. and Slud, S. (1988) Central limit theorems for nonlinear functionals of stationary Gaussian processes. Probab. Th. Rel. Fields, **80**, 323-346.

Chambers, M.J. (1993) The estimation of continuous parameter long-memory time series models. preprint.

Chan, N.H. and Wei, C.Z. (1988) Limiting distributions of least squares estimates of unstable autoregressive processes. Ann. Statist., **16**, 367-401.

Chatterjee, S. and Yilmaz, M.R. (1992a) Chaos, fractals and statistics. (with discussion). Statist. Sci., **7**, 49-68.

Chatterjee, S. and Yilmaz, M.R. (1992b) Reply to comments on "Chaos, fractals and statistics." (with discussion). *Statist. Sci.*, **7**, 114-117.

Chatterjee, S. and Yilmaz, M. (1992c) Use of estimated fractal dimension in model identification for time series. *J. Statist. Comput. Simul.*, **41**, 129-141.

Chayes, J.T. Chayes, L., Grannan, E., and G. Swindle, G. (1991) Phase transitions in Mandelbrot's percolation process in three dimensions. *Probab. Th. Rel. Fields*, **90**, 291-300.

Chen, S.K. (1988) FDSM: A program for a fractional-differencing seasonal model. *Am. Statist.*, **42**, 226-226.

Cheng, B. and Robinson, P.M. (1991) Density estimation in strongly dependent non-linear time series. *Statistica Sinica*, **1**, 335-359.

Cheng, B. and Robinson, P.M. (1992) Semiparametric estimation of time series with strong dependence. *preprint*.

Cheng, C.-S. (1983) Construction of optimal balanced incomplete block designs for correlated observations. *Ann. Statist.*, **11**, 240-246.

Cheung, Y.W. (1990) Long memory in foreign exchange rates and sampling properties of some statistical procedures to long memory series. *PhD dissertation, University of Pennsylvania*.

Cheung, Y.W. (1993) Long memory in foreign exchange rates. *J. Bus. Econ. Statist.*, **11**, 93-101.

Cheung, Y.-W. and Diebold, F.X. (1990) On maximum-likelihood estimation of the differencing parameter of fractionally integrated noise with unknown mean. *Discussion paper, Institute for Empirical Macroeconomics, Federal Reserve Bank of Minneapolis*.

Cheung, Y.-W. and Lai, K.S. (1993) A fractional cointegration analysis of purchasing power parity. *J. Bus. and Econ. Statist.*, **11**, 103-112.

Church, K.W. and Helfman, J.I. (1992) Dotplot: A program for exploring self-similarity in millions of lines of text and code. *Proc. Comp. Sci. and Statist., 9th Annual Symp. on the Interface, UCLA, Vol. 24*, pp. 58-67.

Cioczek-Georges, R. and Taqqu, M.S. (1992) Does asymptotic linearity of the regression extend to stable domains of attraction ? *preprint*.

Clarke, B.R. (1983) Uniqueness and Fréchet differentiability of functional solutions to maximum likelihood type equations. *Ann. Statist.*, **11**, 1196-1205.

Cline, D. (1983) Estimation and linear prediction for regression, autoregression and ARMA with infinite variance data. *Ph.D. thesis, Department of Statistics, Colorado State University*.

Cochrane, D. and Orcutt, G.H. (1949) Application of least squares regression to relationships containing autocorrelated error terms. *J. Am. Statist. Assoc.*, **44**, 32-61.

Constantine, A.G. and Hall, P. (1994) Characterizing surface smoothness via estimation of effective fractal dimension. *J. Roy. Statist Soc., Series B*, **56, No. 1**, 97-113.

Cooley, J.W., Lewis, P.A.W., and Welch, P.D. (1970) The application of the fast Fourier transform algorithm to the estimation of spectra and cross-spectra. *J. Sound Vib.*, **12**, 339-352.

Cootner, P.H. (ed.) (1964) The random character of stock market prices. *MIT Press, Cambridge, Mass.*

Coster, M. and Chermant, J.L. (1983) Recent developments in quantitative fractography. *Int. Met. Rev.*, **28**, 234-238.

Cox, D.R. (1974) Discussion on "Stochastic modelling of riverflow time series" (by A.J. Lawrence and N.T. Kottegoda). *J. Roy. Statist. Soc.*, *Series A*, **140**, 34.

Cox, D.R. (1984) Long-range dependence: A review. *Statistics: An Appraisal. Proceedings 50th Anniversary Conference. H.A. David, H.T. David (eds.). The Iowa State University Press*, pp. 55-74.

Cox, D.R. (1991) Long-range dependence, non-linearity and time irreversibility. *J. Time Ser. Anal.*, **12, No. 4**, 329-335.

Cox, D.R. and Townsend, M.W.H. (1947) The use of the correlogram in measuring yarn irregularity. *Proc. Int. Wool Textile Organization, Technical Committee 2*, pp. 28-34.

Csörgö, M. and Horvath, L. (1988) Nonparametric methods for changepoint problems. *Hanbook of Statistics, Vol. 7, P.R. Krishnaiah, C.R. Rao (eds.), North Holland, Elsevier, Amsterdam*, pp. 403-425.

Curl, R.L. (1986) Fractal dimensions and geometries of caves. *Math. Geol.*, **18**, 765-783.

Cutler, C.D. (1992) Comment on "Chaos, fractals and statistics". *Statist. Sci.*, **7**, 91-94.

D'Agostino, R.B. and Stephens, M.A. (1986) Goodness-of-fit techniques. *Marcel Dekker, New York.*

Dahlhaus, R. (1987) Discussion of session on self-similar processes. *Proceedings of the 46th Session of ISI, Tokyo, Book 4*, pp. 258-260.

Dahlhaus, R. (1989) Efficient parameter estimation for self-similar processes. *Ann. Statist.*, **17**, 1749-1766.

Dahlhaus, R. (1991) Efficient location and regression estimation for long-range dependent regression models. *preprint.*

Damerau, F.J. and Mandelbrot, B.B. (1973) Tests of the degree of word clustering in samples of written English. *Linguistics*, **102**, 58-75.

David, H.A. (1985) Bias of s^2 under dependence. *Am. Statist.*, **39**, 201.

Davies, R.B. and Harte, D.S. (1987) Tests for Hurst effect. *Biometrika*, **74**, 95-102.

Davies, N. and Newbold, P. (1979) Some power studies of a portmanteau test of time series model specification. *Biometrika*, **66**, 153-155.

Davies, N., Triggs, C.M., and Newbold, P. (1977) Significance levels of the Box-Pierce portmanteau statistic in finite samples. *Biometrika*, **64**, 517-522.

Davis, R.A., Knight, K., and Liu, J. (1992) M-estimation for autoregression with infinite variance. *Stoch. Proc. Appl.*, **40**, 145-180.

Davis, R.A. and Resnick, S. (1985) Limit theory for moving averages of random variables with regularly varying tail probabilities. *Ann. of Probab.*, **13**, 179-195.

Davis, R.A. and Resnick, S. (1986a) Limit theory for the sample covariance and correlation function of moving averages. *Ann. Statist.*, **14**, 533-558.

Davis, R.A. and Resnick, S. (1986b) Limit theory for the sample correlation function of moving averages. Dependence in Probability and Statistics, E. Eberlein and M.S. Taqqu (eds.), *Birkhäuser, Boston.*

Davison, A.C. and Cox, D.R. (1989) Some simple properties of sums of random variables having long-range dependence. *Proc. Roy. Soc. London, Series A*, **424**, 255-262.

Davydov, Y.A. (1970) The invariance principle for stationary processes. *Theor. Probab. Appl.*, **15**, 487-498.

Deeter, J.E, Boynton, P.E., and Percival, D.B. (1981) Vela X-1 timing experiment. *Progress report, Dept. of Astronomy, University of Washington, Seattle.*

De Geer, G. (1940) Geochronologica Suecica Principles. *Almquist and Wiksells, Stockholm.*

de Haan, L. (1970) On regular variation and its application to weak convergence of sample extremes. *Math. Centre Tracts, 32, Math. Centre, Amsterdam.*

Dehling, H. (1986) Almost sure approximation for U-statistics. *Dependence in Probability and Statistics, E. Eberlein and M.S. Taqqu, (eds.), Birkhäuser, Boston*, pp. 119-136.

Dehling, H. and Taqqu, M.S. (1988) The functional law of the iterated logarithm for the empirical process of some long-range dependent sequences. *Statist. Probab. Lett.*, **7**, 81-85.

Dehling, H. and Taqqu, M.S. (1989) The empirical process of some long-range dependent sequences with application to U-statistics. *Ann. Statist.*, **17**, 1767-1783.

Dehling, H. and Taqqu, M.S. (1991) Bivariate symmetric statistics of long-range dependent observations. *J. Statist. Pl. Inf.*, **28**, 153-165.

De Matteis, A. and Pagnutti, S. (1988) Parallelization of random number generators and long-range correlations. *Numer. Math.*, **53**, 595-608.

Dempster, A.P. and Hwang, J.-S. (1992) Comment on "Statistical methods for data with long-range dependence." *Statist. Sci.*, **7, No. 4**, 416-420.

Devaney, R.L. (1990) Chaos, fractals, and dynamics. *Addison Wesley, California.*

Diebold, F.X. and Nason, J.A. (1990) Nonparametric exchange rate prediction ? *J. Int. Econ.*, **28**, 315-332.

Diebold, F.X. and Rudebush, G.D. (1989) Long memory and persistence in aggregate output. *J. Monet. Econ.*, **24**, 189-209.

Diebold, F.X. and Rudebush, G.D. (1991) On the power of the Dickey-Fuller test against fractional alternatives. *Econ. Lett.*, **35**, 155-160.

Diggle, P. (1990) Time series: A biostatistical introduction. *Oxford University Press, Oxford.*

Dobrushin, R.L. (1979) Gaussian and their subordinated self-similar random fields. *Ann. Probab.*, **7**, 1-28.

Dobrushin, R.L. and Major, P. (1979) Non-central limit theorems for non-linear functionals of Gaussian fields. *Z. Wahr. verw. Geb.*, **50**, 27-52.

Doukhan, P and Leon, J.R. (1991) Estimation du spectre d'un processus Gaussien forment dependant. *C.R. Acad. Sci., Ser. I*, 313, 523-526.

Durbin, J. (1960) The fitting of time series models. *Int. Statist. Rev.*, **28**, 233-244.

Durbin, J. and Watson, G.S. (1950) Testing for serial correlation in least squares regression. *Biometrika*, **37**, 409-428.

Eberhard, J.W. and Horn, P.M. (1978) Excess 1/f noise in metals. *Phys. Rev. B*, **18**, 6681-6693.

Eberlein, E. and Taqqu, M.S. (eds.) (1986) Dependence in Probability and Statistics. *Birkhäuser, Boston.*

Eichenauer-Herrmann, J. and Grothe, H. (1989) A remark on long-range correlations in multiplicative congruential pseudo random number generators. *Numer. Math.*, **56**, 609-611.

Eicker, F. (1967) Limit theorems for regression with unequal and dependent errors. *Proc. Fifth Berkely Symp. on Math. Statist. and Probab., Vol. 1, University of California, Berkely*, pp. 59-82.

Engel, C. and Hamilton, J.D. (1990) Long swings in the dollar: Are they in the data and do markets know it ? *American Economic Review*, **80**, 689-713.

Engle, R.F. and Granger, C.W.J. (1987) Co-integration and error correction: Representation, estimation and testing. *Econometrica*, **55**, 251-276.

Epps, T.W. (1988) Testing that a Gaussian process is stationary. *Ann. Statist.*, **16**, 1667-1683.

Erramilli, A. and Singh, R.P. (1992) An application of deterministic chaotic maps to characterize packet traffic. *preprint.*

Erramilli, A. and Willinger, W. (1993) Fractal properties of packet traffic measurements. *preprint.*

Falconer, K.J. (1985) The geometry of fractal sets. *Cambridge University Press, Cambridge.*

Falconer, K.J. (1986) Random fractals. *Math. Proc. Cambridge Philos. Soc.*, **100**, 559-582.

Fama, E.F. (1965) The behavior of stock market prices. *J. Bus.*, **38**, 34-105.

Fama, E.F. (1966) Filter rules and stock market trading. *J. Bus., Supplement, January 1966*, **29, No. 1, Part II**.

Feller, W. (1951) The asymptotic distribution of the range of sums of independent random variables. *Ann. Math. Statist.*, **22**, 427-432.

Fiering, M. (1967) Streamflow synthesis. *Harvard University Press, Cambridge.*

Fisher, R.A. (1925) Statistical methods for research workers. *Oliver and Boyd, Edinburgh.*

Flandrin, P. (1989) On the spectrum of fractional Brownian motions. *IEEE Trans. Inf. Th.*, **35**, 197-199.

Flandrin, P. (1992) Wavelet analysis and synthesis of fractional Brownian motion. *IEEE Trans. Inf. Th.*, **38**, 910-917.

Fowler, H.J. and Leland, W.E. (1991) Local area network traffic characteristics, with implications for broadband network congestion management. *IEEE J. Sel. Areas Commun.*, **9**, 1139-1149.

Fox, R. and Taqqu, M.S. (1985) Non-central limit theorems for quadratic forms in random variables having long-range dependence. *Ann. Probab.*, **13**, 428-446.

Fox, R. and Taqqu, M.S. (1986) Large sample properties of parameter estimates for strongly dependent stationary Gaussian time series. *Ann. Statist.*, **14**, 517-532.

Fox, R. and Taqqu, M.S. (1987) Central limit theorems for quadratic forms in random variables having long-range dependence. *Probab. Th. Rel. Fields*, **74**, 213-240.

Friedlander, S.K. and Topper, L. (1961) Turbulence: Classic papers on statistical theory. *Interscience, New York.*

Fristedt, B. (1974) Sample functions of stochastic processes with stationary independent increments. *Advances in Probability and Related Topics, Vol. 3, P. Ney and S. Port, Dekker*, 241-396.

Fröhlich, J. (editor) (1983) Scaling and self-similarity in physics: Renormalization in statistical mechanics and dynamics. *Progress in Physics, Vol. 7, Birkhäuser, Boston.*

Fujikawa, H. and Matsushita, M. (1990) Fractal growth of bacterial colonies (Japanese). *Proc. Inst. Statist. Math.*, **38**, 61-72.

Gadrich, T. (1993) Parameter estimation for ARMA processes with symmetric stable innovations and related problems. *Ph.D. thesis, Technion.*

Gadrich, T. and Adler, R.J. (1993) Parameter estimation for ARMA processes with symmetric stable innovations. *preprint.*

Garrett, M.W., Latouche, G., Sherman, R., and Willinger, W. (1993) Fast methods for generating long sequences of data with long-range dependence. *preprint.*

Garrett, M.W. and Willinger, W. (1993) Statistical analysis of a long sample of variable bit rate video traffic. *preprint.*

Gastwirth, J.L. and Rubin, H. (1967) The effect of autoregressive dependence on a non-parametric test. *IEEE Trans. Inf. Th.*, **IT-13**, 311-313.

Gastwirth, J.L. and Rubin, H. (1975a) The asymptotic distribution theory of the empiric CDF for mixing stochastic processes. *Ann. Statist.*,

3, 809-824.

Gastwirth, J.L. and Rubin, H. (1975b) The behavior of robust estimators on dependent data. *Ann. Statist.*, **3**, 1070-1100.

Gay, R. and Heyde, C.C. (1990) On a class of random field models which allows long range dependence. *Biometrika*, **77**, 401-403.

Gelfand, I.M. and Vilenkin, N.Ya. (1964) Generalized functions. *Academic Press, New York.*

Geweke, J. (1992) Comment on "Chaos, fractals and statistics". *Statist. Sci.*, **7**, 94-101.

Geweke, J. and Porter-Hudak, S. (1983) The estimation and application of long memory time series models. *J. Time Ser. Anal.*, **4**, 221-237.

Ghosh, S. (1987) Some new tests of normality using methods based on transforms. *PhD thesis, Dept. of Statistics, University of Toronto.*

Ghosh, S. (1993) A new graphical tool to detect non-normality. *preprint.*

Ghosh, S. and Ruymgaart, F. (1992) Application of empirical characteristic functions in some multivariate problems. *Canad. J. Statist.*, **20**, 429-440.

Gilbertson, L.N. and Zipp, R.D. (1981) Fractography and material Science. *Am. Soc. Testing Materials, Philadelphia.*

Giraitis, L. (1983) Convergence of some nonlinear transformations of a Gaussian sequence to self-similar processes (Russian). *Litovsk. Mat. Sb.*, **23, No. 1**, 58-68.

Giraitis, L. (1985) Central limit theorems for functionals of a linear process. *Litovskii Matematickii Sbornik*, **25**, 43-57.

Giraitis, L. (1989) Central limit theorems for polynomial forms. *Lietuvos Matematikos Rinkinys*, **29**, 266-289.

Giraitis, L. and Leipus, B. (1991) Functional CLT for nonparametric estimates of the spectrum and change-point problem for a spectral function (in Russian). *Litovskii Matematicheskii Sbornik*, **30, No. 4**, pp. 674-697.

Giraitis, L. and Leipus, B. (1992) Testing and estimating in the change-point problem of the spectral function. *Lith. Math. J.*, **32, No. 1**, 1-19.

Giraitis, L. and Surgailis, D. (1985) Central limit theorems and other limit theorems for functionals of Gaussian processes. *Z. Wahrsch. verw. Geb.*, **70**, 191-212.

Giraitis, L. and Surgailis, D. (1986) Multivariate Appell polynomials and the central limit theorem. *Dependence in Probability and Statistics, E. Eberlein and M.S. Taqqu (eds.). Birkhäuser, Boston*, pp. 21-72.

Giraitis, L. and Surgailis, D. (1989) A limit theorem for polynomials and the central limit theorem. *Lietuvos Matematikos Rinkinys*, **29**, 290-311.

Giraitis L. and Surgailis D. (1990a) A central limit theorem for quadratic forms in strongly dependent linear variables and application to asymptotical normality of Whittle's estimate. *Probab. Th. Rel. Fields*, **86**,

87-104.

Giraitis, L. and Surgailis, D. (1990b) On shot noise processes with long range dependence. *Prob. Th. and Math. Statist., Proc. 5th Vilnius Conf., Vol. 1, B. Grigelionis, Yu, V. Prohorov, V.V. Sazonov, V.Statulevivius (eds.), VSP, Netherlands, Vol. 1*, pp. 401-408.

Giraitis, L. and Surgailis, D. (1991) On shot noise processes attracted to fractional Levy motion. *Stable Processes and Related Topics, S. Cambanis, G. Samorodnitsky and M.S. Taqqu (eds.), Birkhäuser, Boston*, 261-273.

Gisselquist, R. (1973) A continuum of collision process limit theorems. *Ann. Probab.*, **1**, 231-239.

Gleser, L.J. and Moore, D.S. (1983) The effect of dependence on chisquared test and empiric distribution tests of fit. *Ann. Statist.*, **11**, 1100-1108.

Godano, C. (1991) Scale invariance for a heterogeneous fault model. *Pure Appl. Geophys.*, **137**, 85-94.

Goldberger, A.L. and West, B.J. (1992) Chaos and order in the human body. *Chance*, **5, No. 1**, 47-55.

Good, I.J. (1983) Fractional dimensions, continued fractions, and Markovity (reconstructed record). *J. Statist. Comput. Simul.*, **18**, 80-84.

Goodman, V. and James Kuelbs, J. (1991) Rates of clustering for some Gaussian self-similar processes. *Probab. Th. Rel. Fields*, **88**, 47-75.

Gorodetskii, V.V. (1977) On the convergence to semi-stable Gaussian processes. *Th. Probab. Appl.*, **22**, 498-508.

Gourieroux, C., Maurel, A., and Monfort, A. (1989) Least squares and fractionally integrated regressors. *preprint*.

Gradshteyn, I.S. and Ryzhik, I.M. (1965) Tables of Integrals, Series and Products. *Academic Press, London*.

Graf H.P. (1983) Long-range correlations and estimation of the selfsimilarity parameter. *PhD thesis, ETH Zürich*.

Graf, H.P., Hampel, F.R., and Tacier, J. (1984) The problem of unsuspected serial correlations. *Robust and Nonlinear Time Series Analysis. J. Franke, W.Härdle, R.D. Martin (eds.), Lecture Notes in Statistics, Vol. 26, Springer, New York*, pp. 127-145.

Graf, S. (1987) Statistically self-similar fractals. *Probab. Th. Rel. Fields*, **74**, 357-392.

Granger, C.W.J. (1966) The typical spectral shape of an economic variable. *Econometrica*, **34**, 150-161.

Granger, C.W.J. (1978) New classes of time series models. *Statistician*, **27**, 237-253.

Granger, C.W.J. (1980) Long memory relationships and the aggregation of dynamic models. *J. Econometrics*, **14**, 227-238.

Granger, C.W.J. (1992) Comment on "Chaos, fractals and statistics." *Statist. Sci.*, **7**, 102-104.

Granger, C.W.J. and Hatanaka, M. (1964) Spectral analysis of economic time series. *Princeton University Press, Princeton.*

Granger, C.W.J. and Joyeux, R. (1980) An introduction to long-range time series models and fractional differencing. *J. Time Ser. Anal.*, **1**, 15-30.

Granger, C.W.J. and Newbold, P. (1977) Forecasting economic time series. *Academic Press, New York.*

Graversen, S.E. and Violle-Apiala, J. (1986) α−self-similar Markov processes. *Probab. Th. Rel. Fields*, **71**, 149-158.

Gray, H.L., Zhang, N., and Woodward, W.A. (1989) On generalized fractional processes. *J. Time Ser. Anal.*, **10**, 233-257.

Greene, M.T. and Fielitz, B.D. (1977) Long-term dependence in common stock returns. *J. Financial Econ.*, **4**, 339-349.

Greene, M.T. and Fielitz, B.D. (1979) The effect of long-term dependence on risk return models of common stocks. *Op. Res.*, **27**, 944-951.

Gregotski, M.E., Olivia Jensen, O., and Arkani-Hamed, J. (1991) Fractal stochastic modeling of aeromagnetic data. *Geophys.*, **56**, 1706-1715.

Grenander, U. (1954) On the estimation of regression coefficients in the case of autocorrelated disturbance. *Ann. Math. Statist.*, **25**, 252-272.

Grenander, U. and Rosenblatt, M. (1957) Analysis of stationary time series. *Wiley, New York.*

Grenander, U. and Szegö, G. (1958) Toeplitz forms and their application. *Univ. of California Press, Berkeley.*

Griffeath. D. (1992) Comment on "Chaos, fractals and statistics." *Statist. Sci.*, **7**, 104-108.

Gupta, S.N. (1988) Parameter estimation in fractionally differenced noise processes. *Metron*, **46**, 227-235.

Gupta, S.N. (1992) Estimation in long memory time series models. *Comm. Statist. A*, **21**, 1327-1338.

Halford, D. (1968) A general mechanical model for $|f|^2$ spectra with special reference to $1/|f|$ flicker noise. *Proc. IEEE*, **56**, 251-258.

Hall P. (1982) On some simple estimates of an exponent of regular variation. *J. Roy. Statist. Soc., Series B*, **44**, 37-42.

Hall, P. (1993) On the performance of box-counting estimators of fractal dimension. *Biometrika*, **80, No. 1**, 246-252.

Hall, P. and Hart, J.D. (1990) Nonparametric regression with long-range dependence. *Stoch. Proc. Appl.*, **36**, 339-351.

Hall, P. and Welsh, A.H. (1984) Best attainable rates of convergence for estimates of parameters of regular variation. *Ann. Statist.*, **12**, 1079-1084.

Hall, P. and Welsh, A.H. (1985) Adaptive estimates of parameters of regular variation. *Ann. Statist.*, **13**, 331-341.

Hambly, B.M. (1992) Brownian motion on a homogeneous random fractal. *Probab. Th. Rel. Fields*, **94**, 1-38.

Hampel, F.R. (1971) A general qualitative definition of robustness. *Ann.*

Math. Statist., **42**, 1887-1896.

Hampel, F.R. (1973) Robust estimation: A condensed partial survey. *Z. Wahrsch. verw. Geb.*, **27**, 87-104.

Hampel, F.R. (1974) The influence curve and its role in robust estimation. *J. Am. Statist. Assoc.*, **69**, 383-393.

Hampel, F.R. (1978) Modern trends in the theory of robustness. *Math. Oper. Statist., Ser. Statist.*, **9**, 425-442.

Hampel, F.R. (1979) Discussion of the meeting on robustness. *Proc. 42nd Session of the ISI, Manila, Book 2*, pp. 100-102.

Hampel F.R. (1987) Data analysis and self-similar processes. *Proceedings of the 46th Session of ISI, Tokyo, Book 4*, pp. 235-254.

Hampel, F.R., Marazzi, A., Ronchetti, E., Rousseeuw, P.J., Stahel, W., and Welsh, R.E. (1982) *Handouts for the Instructional Meeting on Robust Statistical Methods. 15th European Meeting of Statisticians, Palermo, Italy. Fachgruppe für Statistik, ETH, Zürich.*

Hampel, F.R., Ronchetti, E.M., Rousseeuw, P.J. and Stahel, W.A. (1986) Robust statistics. The approach based on influence functions. *Wiley, New York.*

Hannan, E. (1963a) Regression for time series. *Time Series Analysis, M. Rosenbaltt (ed.), Wiley, New York*, pp. 17-37.

Hannan, E. (1963b) Regression for time series with errors of measurements. *Biometrika*, **50**, 293-302.

Hannan, E. (1970) Multiple time series. *Wiley, New York.*

Hannan, E. (1976) The asymptotic distribution of serial covariances. *Ann. Statist.*, **4**, 366-399.

Hannan, E. (1980) The estimation of the order of an ARMA process. *Ann. Statist.*, **8**, 1071-1081.

Hannan, E. and Kanter, M. (1977) Autoregressive processes with infinite variance. *J. Appl. Probab.*, **10**, 411-415.

Hara, H. and Koyama, J. (1991) Activations of random system and its power distribution (Japanese). *Proc. Inst. Statist. Math.*, **39**, 73- 83.

Hardin Jr., C.D. (1982) On the spectral representation of symmetric stable processes. *J. Multivar. Anal.*, **12**, 385-401.

Hardin Jr., C.D., Samorodnitsky, G., and Taqqu, M.S. (1991) Non-linear regression of stable random variables. *Ann. Probab.*, **1**, 582-612.

Haslett, J. and Raftery, A.E. (1989) Space-time modelling with long-memory dependence: Assessing Ireland's wind power resource. Invited paper with discussion. *J. Appl. Statist.*, **38**, 1-50.

Hassler, U. (1993a) Regression of spectral estimators with fractionally integrated time series. *preprint.*

Hassler, U. (1993b) Unit root test: the autoregressive approach in comparison with the periodogram regression. *Statistische Hefte*, 67-82.

Hattori, K. and Hattori, T. (1991) Self-avoiding process on the Sierpinski gasket. *Probab. Th. Rel. Fields*, **88**, 405-428.

Haubrich, J. (1989) Consumption and fractional differencing: Old and new anomalies. *preprint*.

Haubrich, J. and Lo, A. (1989) The sources and nature of long-range dependence in the business cycle. *NBER Working Paper No. 2951*.

Hayward, J., Orford, J.D., and Whalley, W.B. (1989) Three implementations of fractal analysis of particle outlines. *Comp. Geosci.*, **15**, 199-207.

Heeke, H. (1991) Statistical multiplexing gain for variable bit rate codecs in ATM networks. *Int. J. Digit. Analog. Commun. Syst.*, **4**, 261-268.

Heeke, H. (1993) A traffic-control algorithm for ATM networks. *IEEE Trans. Circuits Syst. Video Technol.*, **3**, 182-189.

Helland, K.N. and Van Atta, C.W. (1978) The "Hurst phenomenon" in grid turbulence. *J. Fluid. Mech.*, **85**, 573-589.

Helms, B., Kaen, F., and Rosenman, R. (1984) Memory in commodity futures contracts. *J. Fut. Markets*, **4**, 559-567.

Hernad, D., Mouillart, M., and Strauss-Kahn, D. (1978) On the good use of R/S (French). *Rev. Statist. Appl.*, **26, No. 4**, 61-77.

Heyde, C.C. (1985) On some new probabilistic developments of significance to statistics: Martingales, long range dependence, fractals, and random fields. *A Celebtration of Statistics: The ISI Centenary Vol., A.C. Atkinson, S.E. Fienberg (eds.), Springer, New York*, pp. 355-368.

Heyde, C.C. (1987) Discussion of session on self-similar processes. *Proceedings of the 46th Session of ISI, Tokyo, Book 4*, p. 255.

Heyman, D., Tabatabai, A., and Lakshman, T.V. (1991) Statistical analysis and simulation of video teleconferencing in ATM networks. *IEEE Trans. Circuits. Syst. Video Technol.*, **2**, 49-59.

Hibbert, D.B. (1991) Fractals in chemistry. *Chemom. Int. Lab. Syst.*, **11**, 1-11.

Higuchi, T. (1989a) Comments on "Long-memory models and their statistical properties" (Japanese). *J. Japn. Stat. S.*, **19**, 241-244.

Higuchi, T. (1989b) Fractal analysis of time series (Japanese). *Proc. Inst. Statist. Math.*, **37**, 209-233.

Hipel, K.W. and McLeod, A.I. (1978a) Preservation of the rescaled adjusted range, Parts 1, 2 and 3. *Water Resources Res.*, **14**, 491-518.

Hipel, K.W. and McLeod, A.I. (1978b) Preservation of the rescaled adjusted range. Water Resources Res., **14**, 509-516.

Hipel, K.W. and McLeod, A.I. (1980) Perspectives in stochastic hydrology. *Time Series, O.D. Anderson (ed.), Elsevier/N. Holland*, pp. 73-102.

Hirata, T. and Imoto, M. (1991) Multifractal analysis of spatial distribution of microearthquakes in the Kanto region. *Geophys. J.*, **107**, 155-162.

Ho, H.-C. (1992) On limiting distributions of nonlinear functions of noisy Gaussian sequences. *Stochast. Ann. Appl.*, **10**, 417-430.

Ho, H.-C. and Sun, T.-C. (1987) A central limit theorem for non-instantaneous filters of a stationary Gaussian process. *J. Multivar.*

Anal., **22**, 144-155.

Hodges, J.L. and Lehmann, E.L. (1970) Deficiency. *Ann. Statist.*, **41**, 783-801.

Holley, R. and Waymire, E.C. (1992) Multifractal dimensions and scaling exponents for strongly bounded random cascades. *Ann. Appl. Probab.*, **2**, 819-845.

Hooghiemstra, G. (1987) On functionals of the adjusted range process. *J. Appl. Probab.*, **24**, 252-257.

Hosking, J.R.M. (1981) Fractional differencing. *Biometrika*, **68**, 165-176.

Hosking, J.R.M. (1982) Some models of persistence in time series. *Times Series Analysis: Theory and Practice 1, North Holland, New York, 641-654*

Hosking, J.R.M. (1984) Modelling persistence in hydrological time series using fractional differencing. *Water Resources Res.*, **20**, 1898-1908.

Hosking, J.R.M. (1985) Fractional differencing modeling in hydrology. *Water Resources Bull.*, **21**, 677-682.

Huber, P. (1964) Robust estimation of a location parameter. *Ann. Math. Statist.*, **35**, 73-101.

Huber, P. (1967) The behavior of maximum likelihood estimates under nonstandard conditions. *Proc. 5th Berkeley Symp. on Math. Statist. and Probab., L. LeCam and J. Neyman (eds.), Univ. of California, Berkeley*, 221-233.

Huber, P. (1981) Robust statistics. *Wiley, New York*.

Hudson, W.N. and Mason, J.D. (1982) Operator-self-similar processes in a finite-dimensional space. *Trans. Am. Math. Soc.*, **273**, 281-297.

Huele, A.F. (1993) Long-memory time series models and fractional differencing. *PhD thesis, University of Amsterdam*.

Hughes, B.D., Montroll, E.W., and Shlesinger, M.F. (1982) Fractal random walks. *J. Statist. Phys.*, **28**, 111-126.

Hughes, B.D., Shlesinger, M.F., and Montroll, E.W. (1981) Random walks with self-similar clusters. *Proc. Natl. Acad. Sci. USA*, **78**, 3287-3291.

Hui, Y.V. and Li, W.K. (1988) On fractionally differenced periodic processes. *Manuscript, Chinese University of Hong Kong and University of Hong Kong*.

Hunt, F. (1990) Error analysis and convergence of capacity dimension algorithms. *SIAM J. Appl. Math.*, **50**, 307-321.

Hurst, H.E. (1951) Long-term storage capacity of reservoirs. *Trans. Am. Soc. Civil Engineers*, **116**, 770-799.

Hurst, H.E. (1955) Methods of using long-term storage in reservoirs. Proc. Inst. Civil Engin., Part I, pp. 519-577.

Hurst, H.E., Black, R.P., and Simaika, Y.M. (1965) Long-term storage: An experimental study. *Constable Press, London*.

Hurvich, C.M. and Beltrao, K.I. (1992) Asymptotics for the low-frequency ordinates of the periodogram of a long-memory time series.

J. Time Ser. Anal., **14**, 455-472.

Hurvich, C.M. and Beltrao, K.I. (1994) Automatic semiparametric estimation of the memory parameter of a long memory time series. *J. Time Ser. Anal.*, **15, No. 3**, 285-302.

Hurvich, C.M. and Ray, B.K. (1994) Estimation of the long-memory parameter for nonstationary or noninvertible fractionally differenced integrated processes. *preprint.*

Hwang, J.-S. (1992) Prototype Bayesian estimation of US employment and unemployment rates. *PhD dissertation, Dept. of Statistics, Harvard University.*

Ibragimov, I.A. and Linnik, J.V. (1971) Independent and stationary sequences of random variables. *Walters-Noordhoff, Groningen.*

Ito, K. (1951) Multiple Wiener integral. *J. Math. Soc. Jap.*, **3**, 157-164.

Ivanov, A.V. and Leonenko, N. (1989) Statistical analysis of random fields. *Kluwer Academic Publishers, Dordrecht.*

Janacek, G.J. (1982) Determining the degree of differencing for time series via log spectrum. *J. Time Ser. Anal.*, **3**, 177-183.

Jeffreys, H. (1939, 1948, 1961) Theory of probability. *Clarendon Press, Oxford.*

Jenkins, G.M. and Chanmugan, J. (1962) The estimation of slope when the errors are autocorrelated. *J. Roy. Statist. Soc., Series B*, **24**, 199-214.

Johansen, S. (1988) Statistical analysis of cointegration vectors. *J. Econ. Dynam. Cont.*, **12**, 231-254.

Johnson, A.R., Milne, B.T., and Wiens, J.A. (1992) Diffusion in fractal landscapes: Simulations and experimental studies of Tenebrionid beetle movements. *Ecology*, **73**, 1968-1983.

Jona-Lasinio, G. (1977) Probabilistic approach to critical behavior. *New Developments in Quantum Field Theory and Statistical Mechanics, Proc. Cargese Summer Institute 1976, M. Lévy and P. Mitter (eds.), Plenum, New York*, pp. 419-446.

Jones, P.D. and Briffa, K.R. (1992) Global surface air temperature variations during the twentieth century: Part 1, spatial, temporal and seasonal details. *The Holocene*, **2**, 165-179.

Jones, J.G., Thomas, R.W., and Earwicker, P.G. (1989) Fractal properties of computer-generated and natural geophysical data. *Comp. Geosci.*, **15**, 227-235.

Jones, J.G., Thomas, R.W., Earwicker, P.G., and Addison, S. (1991) Multiresolution statistical analysis of computer-generated fractal imagery. *CVGIP: Graph. Mod. Im. Proc.*, **53**, 349-363.

Journel, A.G. and Huijbregts, C.J. (1978) Mining geostatistics. *Academic Press, London.*

Kaen, F.R. and Rosenman, R.E. (1986) Predictable behavior in financial markets: Some evidence in support of Heiner's hypothesis. *Am. Econ. Rev.*, **76**, 212-220.

Kaneshige, I. (1960) Measurement of power spectra of roadways. *Isuzu Motor Company Tech. Rep. No. 3379 (in Japanese)*.

Kano, F. Abe, I., Kamaya, H., and Ueda, I. (1991) Derivation of Langmuir and Freundlich adsorption isotherms (Japanese). *Proc. Inst. Statist. Math.*, **39**, 53-61.

Kashahara, M. and Maejima, M. (1988) Weighted sums of i.i.d. random variables attracted to integrals of stable processes. *Probab. Th. Rel. Fields*, **78, No. 1**, 75-96.

Kashahara, M., Maejima, M., and Vervaat, W. (1988) Log-fractional stable processes. *Stoc. Proc. Appl.*, **30, No. 2**, 329-339.

Kashyap, R.L. and Eom, K.-B. (1988) Estimation in long-memory time series model. *J. Time Ser. Anal.*, **9**, 35-41.

Kaye, B.H. (1989) A random walk through fractal dimensions. *VCH Publ. New York*.

Kesten, H. and Spitzer, F. (1979) A limit theorem related to a new class of self-similar processes. *Z. Wahrsch. verw. Geb.*, **50**, 5-25.

Keynes, J.M. (1940) On the method of statistical business research. *Econ. J.*, **L**, 145-156.

Kiefer, J. and Wynn, H.P. (1981) Optimum balanced block and latin square designs for correlated observations. *Ann. Statist.*, **9**, 737-757.

Kiefer, J. and Wynn, H.P. (1984) Optimum and minimax exact treatment designs for one-dimensional autoregressive error processes. *Ann. Statist.*, **9**, 737-757.

Kiu, S.W. (1980) Semi-stable Markov processes in R^n. *Stoch. Proc. Appl.*, **10**, 183-191.

Klemes V. (1974) The Hurst phenomenon: A puzzle? *Water Resources Res.*, **10**, 675-688.

Klemes V. (1978) Physically based stochastic hydrologic analysis. *Adv. Hydroscience*, **11**, 285-356.

Klemes V. (1982) Empirical and causal models in hydrology. *Studies in Geophysics. National Academy Press, Washington, D.C.*, 95-104.

Klemes, V., Srikanthan, R., and McMahon, T.A. (1981) Long-memory flow models in reservoir analysis: What is their practical value? *Water Resources Res.*, **17**, 737-751.

Klüppelberg, C. and Mikosh, T. (1991) Spectral estimates and stable processes. *preprint*.

Klüppelberg, C. and Mikosh, T. (1992) Parameter estimation for ARMA models with infinite variance innovations. *preprint*.

Kohmoto, M. (1989) Statistical mechanics formalism for multifractals (Japanese). *Proc. Inst. Statist. Math.*, **37**, 63-70.

Kokoszka, P.S. and Taqqu, M.S. (1992) Asymptotic dependence of stable self-similar processes of Chentsov type. *Probability in Banach Spaces, Proc. 8th Int. Conf., M.G. Hahn, R.M. DUdley and J. Kuelbs (eds.). Birkhäuser, Boston*, pp. 152-165.

Kokoszka, P.S. and Taqqu, M.S. (1993a) Asymptotic dependence of moving average type self-similar stable random fields. *preprint*.

Kokoszka, P.S. and Taqqu, M.S. (1993b) New classes of self-similar symmetric stable random fields. *preprint*.

Kokoszka, P.S. and Taqqu, M.S. (1993c) Fractional ARIMA with stable innovations. *preprint*.

Kolmogorov, A.N. (1940) Wienersche Spiralen und einige andere interessante Kurven in Hilbertschen Raum. *Comptes Rendus (Doklady) Acad. Sci. URSS (N.S.)* **26**, 115-118.

Kolmogorov, A.N. (1941) Local structure of turbulence in fluid for very large Reynolds numbers. *Transl. in Turbulence. S.K.Friedlander and L.Topper (eds.), 1961, Interscience Publishers, New York*, pp. 151-155.

Kono, N. (1983) Iterated log type strong limit theorems for self-similar processes. *Proc. Japn. Acad. Ser. A*, **59**, 85-87.

Kono, N. (1986) Hausdorff dimension of sample paths for self-similar processes. *Dependence in Probability and Statistics, E.Eberlein, M.S.Taqqu (eds.). Birkhäuser, Boston*, pp. 109-118.

Kono, N. and Maejima, M. (1990) Self-similar stable processes with stationary increments. *Stable Processes and Related Topics, S. Cambanis, G. Samorodnitsky and M.S. Taqqu (eds.), Birkhäuser, Boston*.

Kottegoda N.T. (1970) Applicability of short-memory model to English river flow data. *J. Inst. Water Engrs.*, **24**, 481-489.

Koul, H.L. (1992) M-estimators in linear models with long range dependent errors. *Statist. Probab. Lett.*, **14**, 153-164.

Krebs, W.B. (1991) A diffusion defined on a fractal state space. *Stoch. Proc. Appl.*, **37**, 199-212.

Künsch, H. (1981) Thermodynamics and statistical analysis of Gaussian random fields. *Z. Wahrsch. verw. Geb.*, **58**, 407-427.

Künsch, H. (1984) Infinitesimal robustness for autoregressive processes. *Ann. Statist.*, **12**, 843-863.

Künsch, H. (1986a) Discrimination between monotonic trends and long-range dependence. *J. Appl. Probab.*, **23**, 1025-1030.

Künsch, H. (1986b) Statistical aspects of self-similar processes. *Invited paper, Proc. First World Congress of the Bernoulli Society, Tashkent, Vol. 1, Yu. Prohorov and V.V. Sazonov (eds.), VNU Science Press, Utrecht*, pp. 67-74.

Künsch, H., Beran, J., and Hampel, F. (1993) Contrasts under long-range correlations. *Ann. Statist.*, **21**, 943-964.

Kusuoka, S. and Yin, Z.X.. (1992) Dirichlet forms on fractals: Poincaré constant and resistance. *Probab. Th. Rel. Fields*, **93**, 169-196.

Laha, R.G. and Rohatgi, V.K. (1982) Operator self-similar stochastic processes in R^d. *Stoch. Proc. Appl.*, **12**, 73-84.

Lalley, S.P. (1990) Travelling salesman with a self-similar itinerary. *Probab. Eng. Inf. Sci.*, **4**, 1-18.

Lamperti, J.W. (1962) Semi-stable stochastic processes. *Trans. Am.*

Math. Soc., **104**, 62-78.

Lamperti, J.W. (1972) Semi-stable Markov processes. *Z. Wahrsch. verw. Geb.*, **22**, 205-225.

Lang, R. and Nguyen, Xuan Xanh (1983) Strongly correlated random fields as observed by a random walker. *Z. Wahrsch. verw. Geb.*, **64**, 327-340.

Lasota, A. (1990) Asymptotic behavior of randomly perturbed discrete time dynamical systems. *Stoch. Meth. in Experimental Sciences, W. Kasprzak and A. Weron (eds.), World Sci. Publ., Singapore*, pp. 293-303.

Lawrance, A.J. and Kotegoda, N.T. (1977) Stochastic modelling of river-flow time series (with discussion). *J. Royal Statist. Soc. A*, **140**, 1-47.

Leland, W.E. (1990) LAN traffic behavior from miliseconds to days. *Proc. ITC 7th specilists seminar, Morristown, NJ*, 6.1.1-6.1.6.

Leland, W.E., Taqqu, M.S., Willinger, W., and Wilson, D.V. (1993) Ethernet traffic is self-similar: Stochastic modelling of packet traffic data. *preprint, Bellcore, Morristown.*

Leland, W.E. and Wilson, D.V. (1991) High time-resolution measurement and analysis of LAN traffic: implications for LAN interconnection. *Proceedings of the IEEE INFOCOM'91, Bal Harbor, FL*, pp. 1360-1366.

Leonenko, N.N. (1990) Limit distributions of measures of sojourns of vector Gaussian fields with strong dependence. *Prob. Th. and Math. Statist., Proc. 5th Vilnius Conf., Vol. 1, B. Grigelionis, Yu, V. Prohorov, V.V. Sazonov, V.Statulevivius (eds.), VSP, Netherlands, Vol. 2*, 79- 89.

Leonenko, N.N. and el'-Bassioni, A., Kh. (1987) Limit theorems for some characteristics of level overshoot by a Gaussian field with strong dependence. *Th. Probab. Math. Statist.*, **35**, 59-64.

Leonenko, N.N. and Portnova, A. (1992) Limit theorem correlogram of Gaussian field with long-range dependence. *Ill-posed problems in natural sciences, TVP Sci. Publ.*

Leonenko, N.N. and Portnova, A. (1993) Limit theorem correlogram of chi-square field with long-range dependence. *preprint.*

Levy, J.B. (1983) High variability and long-range dependence. *MS thesis, Cornell University.*

Levy, J. and Taqqu, M.S. (1987) On renewal processes having stable inter-renewal intervals and stable rewards. *Ann. Sci. Math. Québec*, **11**, 97-110.

Levy, J. and Taqqu, M.S. (1991) A characterization of the asymptotic behavior of stationary stable processes. *Stable Processes and Related Topics, G. Samorodnitsky, S. Cambanis and M.S. Taqqu (eds.). Progress in Probability, Vol. 25, Birkhäuser, Boston*, pp. 181-198.

Li, W.K. and McLeod, A.I. (1986) Fractional time series modelling. *Biometrika*, **73**, 217-221.

Li, W.K., Wong, H., and Hui, Y.V. (1985) Fractional seasonal modelling: An example. *ASA Proc. Bus. Econ. Statist. Sect.*, 478-481.

Li, W.K., Wong, H., and Hui, Y.V. (1987) Fractional modelling on emergency room attendance data. *J. Appl. Statist.*, **14**, 239-249.

Lloyd, E.H. (1967) Stochastic reservoir theory. *Adv. Hydrosci.*, *4*, 281.

Lo, A.W. (1991) Long-term memory in stock market prices. *Econometrica*, **59**, 1279-1313.

Logan, B.F., Mallows, C., Rice, S.O., and Shepp, L.A. (1973) Limit distributions of self-normalized sums. *Ann. Probab.*, **1**, 788-809.

Lomnicki, Z. and Zaremba, S.K. (1957) On the estimation of autocorrelations in time series. *Ann. Math. Statist.*, **28**, 1.

Longley, P.A. and Michael Batty, M. (1989) Fractal measurement and line generalization. *Comp. Geosci.*, **15**, 167-183.

Lou, J.H. (1985) Some properties of a special class of self-similar processes. *Z. Wahrsch. verw. Geb.*, **68**, 493-502.

Lovejoy, S. (1982) Area-perimeter relation for rain cloud areas. *Science*, **216**, 185-187.

Lovejoy, S. and Schertzer, D. (1986) Scale invariance, symmetries, fractals and stochastic simulations of atmospheric phenomena. *Bull. Am Meteorol. Soc.*, **67**, 21-32.

Maejima, M. (1981) Some sojourn time problems for strongly dependent Gaussian processes. *Z. Wahrsch. verw. Geb.*, **57**, 1-14.

Maejima, M. (1982) Some limit theorems for sojourn times of strongly dependent Gaussian processes. *Z. Wahrsch. verw. Geb.*, **60**, 359-380.

Maejima, M. (1983a) On a class of self-similar processes. *Z. Wahrsch. Verw. Geb.*, **62, No. 2**, 235-245.

Maejima, M. (1983b) A self-similar process with nowhere bounded sample paths. *Z. Wahrsch. Verw. Geb.*, **65, No. 1**, 115-119.

Maejima, M. (1986a) Some sojourn time problems for two-dimensional Gaussian processes. *J. Multivar. Anal.*, **18**, 15-69.

Maejima, M. (1986b) Sojurns of multidimensional Gaussian processes. *Dependence in Probability and Statistics, E. Eberlein and M.S. Taqqu (eds.), Birkhäuser, Boston*, pp. 91-108.

Maejima, M. (1986c) A remark on self-similar processes with stationary increments. *Can. J. Statist.*, **14**, 81-82.

Maejima, M. (1989a) Self-similar processes and limit theorems. *Sugaku Expositions*, **2**, 103-123.

Maejima, M. (1989a) Comments on "Long-memory models and their statistical properties" (Japanese). *J. Japn. Statist. S.*, **19**, p. 241.

Maejima, M. and Rachev, S.T. (1987) An ideal metric and the rate of convergence to a self-similar process. *Ann. Probab.*, **15**, 708-727.

Major, P.J. (1981a) Multiple Wiener-Ito integrals. *Springer Lecture Notes in Mathematics, 849, Springer, New York.*

Major, P.J. (1981b) Limit theorems for non-linear functionals of Gaussian sequences. *Z. Wahr. verw. Geb.*, **57**, 129-158.

Major, P.J. (1982) On renormalizing Gaussian fields. *Z. Wahrsch. verw. Geb.*, **59**, 515-533.

Mallows, C.L. (1973) Some comments on C_p. *Technometrics*, **15**, 661-675.

Mandelbrot, B.B. (1963a) New methods in statistical economy. *J. Pol. Sci.*, **LXXI, No. 5**, 421-440.

Mandelbrot, B.B. (1963b) The variation of certain speculative prices. *J. Bus. Univ. Chic.*, **36**, 394.

Mandelbrot, B.B. (1965a) Self-similar error clusters in communication systems and the concept of conditional stationarity. *IEEE Trans. Commun. Techn.*, **COM-13**, 71-90.

Mandelbrot, B.B. (1965b) Une classe de processus stochastiques homothétiques à soi; application à la loi climaologique de H.E.Hurst. *Comptes Rendus Acad. Sci. Paris*, **260**, 3274-3277.

Mandelbrot, B.B. (1966) Forecasts of future prices, unbiased markets and 'martingale' models. *J. Bus. Univ. Chic.*, **39**, 242-255.

Mandelbrot, B.B. (1967a) Some noises with $1/f$ spectrum, a bridge between direct current and white noise. *IEEE Trans. Inf. Th.*, **13**, 289-298.

Mandelbrot, B.B. (1967b) Sporadic random functions and conditional spectral analysis; self-similar examples and limits. *Proc. Fifth Berkeley Symp. on Math. Statist. and Prob., Vol. III, Univ. of California Press, Berkeley*, pp. 155-179.

Mandelbrot, B.B. (1967c) How long is the coast of Britain? *Science*, **155**, 636.

Mandelbrot, B.B. (1967d) Sur l'épistemologie de hasard dans les sciences sociales: invariance des lois et vérification des hypothéses. *Encyclopedia de la Pleiade: Logique et connaissance scientifique, J. Piaget (editor), Gallimard, Paris.*

Mandelbrot, B.B. (1969) Long-run linearity, locally Gaussian process, H-spectra and infinite variance. *Int. Econ. Rev.*, **10**, 82-113.

Mandelbrot, B.B. (1971a) When can price be arbitraged efficiently ? A limit to the validity of the random walk and martingale models. *Rev. Econ. Statist.*, **LIII**, 225-236.

Mandelbrot B.B. (1971b) A fast fractional Gaussian noise generator. *Water Resources Res.*, **7**, 543-553.

Mandelbrot, B.B. (1972) Statistical Methodology for nonperiodic cycles: From the covariance to R/S analysis. *Ann. Econ. Soc. Meas.*, **1, No.3**, 259-290.

Mandelbrot, B.B. (1973a) Le problème de la réalité des cycle lents et le "syndrome de Joseph". *Econ. Appl.*, **26**, 349-365.

Mandelbrot, B.B. (1973b) Le syndrom de la variance infinie et ses rapports avec la discontinuitée des prix. *Econ. Appl.*, **26**, 321-348.

Mandelbrot B.B. (1975) Limit theorems of the self-normalized range for weakly and strongly dependent processes. *Z. Wahr. verw. Geb.*, **31**,

271-285.

Mandelbrot, B.B. (1977) Fractals: Form, chance and dimension. *Freeman, San Francisco*.

Mandelbrot, B.B. (1978) Random beadsets, and an alternative to self avoiding random walk: Discrete and fractal stringers (French). *C.R. Acad. Sci. Paris, Vie Acad.*, **286**, 933-936.

Mandelbrot, B.B. (1983a) The fractal geometry of nature. *Freeman, San Francisco*.

Mandelbrot, B.B. (1983b) Fractals. *Encyclopedia of Statistical Sciences, S. Kotz and N. Johnson, Vol. 3, Wiley, New York*, pp. 185-186.

Mandelbrot, B.B. (1983b) Fractional Brownian motions and fractional Gaussian noises. *Encyclopedia of Statistical Sciences, S. Kotz and N. Johnson, Vol. 3, Wiley, New York*, pp. 186-189.

Mandelbrot, B.B. and McCamy, K. (1970) On the secular pole motion and the Chandler wobble. *Geophys. J. Roy. Astron Soc.*, **21**, 217-232.

Mandelbrot, B.B., Passoja, D.E., and Paullay, A.J. (1984) Fractal character of fracture surfaces of metals. *Nature*, **308**, 721-722.

Mandelbrot, B.B. and Taqqu, M.S. (1979) Robust R/S analysis of long run serial correlation. *Proc. 42nd Session of the ISI, Manila, Book 2*, pp. 69-99.

Mandelbrot, B.B. and van Ness, J.W. (1968) Fractional Brownian motions, fractional noises and applications. *SIAM Rev.*, **10, No.4**, 422-437.

Mandelbrot, B.B. and Wallis, J.R. (1968a) Noah, Joseph and operational hydrology. *Water Resources Res.*, **4, No. 5**, 909-918.

Mandelbrot, B.B. and Wallis, J.R. (1968b) Robustness of the rescaled range R/S and the measurement of non-cyclic long-run statistical dependence. *Water Resources Res.*, **5**, 967-988.

Mandelbrot, B.B. and Wallis, J.R. (1969a) Computer experiments with fractional Gaussian noises. *Water Resources Res.*, **5, No.1**, 228-267.

Mandelbrot, B.B. and Wallis, J.R. (1969b) Some long-run properties of geophysical records. *Water Resources Res.*, **5**, 321-340.

Mandelbrot, B.B. and Wallis, J.R. (1969c) Robustness of the rescaled range R/S in the measurement of noncyclic long run statistical dependence. *Water Resour. Res.*, **5**, 967-988.

Mann, J.A., Rains, E.M., and Woyczynski, W.A. (1991) Measuring the roughness of interfaces. *Chemom. Int. Lab. Syst.*, **12**, 169- 180.

Manoukian, E. (1983) Renormalization. *Academic Press, New York*.

Marcus, D. (1983) Non-stable laws with all projections stable. *Z. Wahrsch. verw. Geb.*, **64**, 139-156.

Mardia, K.V. (1980) Tests of univariate and multivariate normality. *Handbook of Statistics, P.R. Krishnaiah (ed.)*, pp. 279-320.

Marinari, E., Parisi, G., Ruelle, D. and Widney, P. (1983) On the interpretation of 1/f noise. *Commun. in Math. Phys.*, **89**, 1-12.

Mark, D.M. and Aronson, P.B. (1984) Scale-dependent fractal dimensions of topographic surfaces: An empirical investigation, with applications in geomorphology and computer mapping. *J. Int. Assoc. Math. Geol.*, **16**, 671-683.

Martin, R.D. and Yohai, V.J. (1986) Influence functionals for time series. Invited paper with discussion. *Ann. Statist.*, **14**, 781-818.

Martin, R.J. (1986) On the design of experiments under spatial correlation. *Biometrika*, **73**, 247-277.

Martin, R.J. and Eccleston, J.A. (1992) A new model for slowly decaying correlations. *Statist. Probab. Lett.*, **13**, 129-145.

Mason, J.D. (1984) A comparison of the properties of operator-stable distributions and operator-self-similar processes. *Limit Theorems in Probability and Statistics, P. Revesz (ed.), Elsevier/N. Holland*, pp. 751- 760.

Masry, E. (1991) Flicker noise and the estimation of the Allan variance. *IEEE Trans. Inf. Th.*, **37**, 1173-1177.

Matalas, N.C. and Wallis, J.R. (1971) Statistical properties of multivariate fractional noise processes. *Water Resources Res.*, **7**, 1460-1468.

Matheron, G. (1962) Traité de Géostatistique Appliquée. *Cambridge Philos. Soc., Tome 1, Technip, Paris*.

Matheron, G. (1973) The intrinsic random functions and their applications. *Adv. Appl. Probab.*, **5**, 439-468.

Mathew, G. and McCormick, W.P. (1989) Complete Poisson convergence result for a strongly dependent stationary Gaussian sequence. *Sankhya A*, **51**, 30-36.

McConnell, T.R. and Taqqu, M.S. (1986) Dyadic approximation of double integrals with respect to symmetric stable processes. *Stoch. Proc. Appl.*, **22**, 323-331.

McCullagh, P. and Nelder, J.A. (1983) Generalized linear models. *Chapman and Hall, London*.

McLeish, D.L. (1975) A maximal inequality and dependent strong laws. *Ann. Probab.*, **3**, 820-839.

Michelson, A.A., Pease, F.G., and Pearson, F. (1935) Measurement of the velocity of light in a partial vacuum. *Contributions from the Mount Wilson Observatory, Carnegie Institution of Washington*, **XXII, No. 522**, 259-294.

Mijnheer, J.L. (1975) Sample path properties of stable processes. *Math. Centre Tracts, 59, Mathematisch Centrum, Amsterdam*.

Milhoj, A. (1981) A test of fit in time series models. *Biometrika*, **68**, 177-188.

Mohr, D. (1981) Estimating the parameter of a fractional Gaussian noise. *Tech. Rep., Dept. of Mathematics, Tulane University, New Orleans*.

Mohr, D. (1981) Modelling data as a fractional Gaussian noise. *PhD thesis, Princeton University*.

Montroll, E.W. and Schlesinger, M.F. (1982) On $1/f$ noise and other distributions with long tails. *Proc. Natl. Acad. Sci. USA*, **79**, 3380-3383.

Moore, D.S. (1982) The effect of dependence on chi-squared tests of fit. *Ann. Statist.*, **4**, 357-369.

Moran, P.A.P. (1964) On the range of cumulative sums. *Ann. Inst. Statist. Math.*, **16**, 109-112.

Moran, P.A.P. (1964) The theory of storage. *Wiley, New York.*

Morgan, J.P. and Chakravarti, I.M. (1988) Block designs for first order and second order neighbour correlations. *Ann. Statist.*, **16**, 1206-1224.

Morgenthaler, S. and Tukey, J.W. (1991) Configural Polysampling. *Wiley, New York.*

Mori, T. and Oodaira, H. (1976) A functional law of the iterated logarithm for sample sequences. *Yokohama Math. J.*, **24**, 35-49.

Mori, T. and Oodaira, H. (1985) The law of the iterated logarithm for self-similar processes represented by multiple Wiener integrals. *Probab. Th. Rel. Fields*, **71**, 367-391.

Mori, T. and Oodaira, H. (1986) The law of the iterated logarithm for self-similar processes represented by multiple Wiener integrals. *Probab. Th. Rel. Fields*, **71**, 367-391.

Mosteller, F. and Tukey, J.W. (1977) Data analysis and regression: A second course in statistics. *Addison-Wesley, Reading, Mass.*

Mountford, T.S. (1992) The critical value for some non-attractive long range nearest particle systems. *Probab. Th. Rel. Fields*, **93**, 67-76.

Munk, W.H. and MacDonald, G.J.F. (1975) The rotation of the earth. *Cambridge University Press, Cambridge.*

Munro, E.H. (1948) Tables of sunspot frequency. *Terrestr. Magnet. and Atm. Electric.*, **55**, 241.

Musha, T. and Higuchi, H. (1976) The $1/f$ fluctuation of a traffic current on an expressway. *Japn. J. Appl. Phys.*, **15**, 1271-1275.

Nagatani, T. (1989) Growth model with phase transition (Japanese). *Proc. Inst. Statist. Math.*, **37**, 199-207.

Nedunuri, S. and Wiseman, N.E. (1987) Displaying random surfaces. *Comput. J.*, **30**, 163-167.

Neumann, G. (1953) On ocean wave spectra and a new method of forecasting wind-generated sea. *U.S. Army Corps of Engineers, Beach Erosion Board, Techn. Memo. No. 43.*

Newbold, P. and Agiakloglou, C. (1993) Bias in the sample autocorrelations of fractional noise. *Biometrika*, **80, No. 3**, 698-702.

Newcomb, S. (1895) Astronomical constants (The elements of the four inner planets and the fundamental constants of astronomy). *Supplement to the American Ephemeris and Nautical Almanac for 1897, US Government Printing Office, Washington DC.*

Newman, C. and Wright, L. (1981) An invariance principle for certain dependent sequences. *Ann. Probab.*, **9**, 671-675.

Nicolis, G., Amellal, A., Dupont, G., and Mareschal, M. (1990) Nonequilibrium states and long range correlations in chemical dynamics. *Stoch. Meth. in Experimental Sciences, W. Kasprzak and A. Weron (eds.), World Sci. Publ., Singapore*, pp. 332-346.

Noakes, D.J., Hipel, K.W., McLeod, A.I., Jimenez, C., and Yakowitz, S. (1988) Forecasting annual geophysical time series. *Int. J. For.*, **4**, 103-115.

Norio, K. (1991) Recent development on random fields and their sample paths. Part III: Self-similar processes. *Soochow J. Math.*, **17**, 327-361.

Nyblom, J. (1991) Testing for constancy of parameters over time. *J. Am. Statist. Assoc.*, **84**, 223-230.

O'Brien, G.L., Torfs, P.J.J.F., and Vervaat, W. (1990) Stationary self-similar extremal processes. *Probab. Th. Rel. Fields*, **87**, 97-119.

O'Brien, G.L. and Vervaat, W. (1983) Marginal distributions of self-similar processes with stationary increments. *Z. Wahrsch. verw. Geb.*, **64**, 129-138.

O'Brien, G.L. and Vervaat, W. (1985) Self-similar processes with stationary increments generated by point processes. *Ann. Probab.*, **13**, 28-52.

Ochiai, M., Ozao, R., Yamazaki, Y., Otsuka, R., and Holz, A. (1990) Self-similarity law of distribution and scaling concept in size reduction of solid (Japanese). *Proc. Inst. Statist. Math.*, **38**, 257-263.

Ogata, Y. (1989) Comments on "Long-memory models and their statistical properties" (Japanese). *J. Japn. Statist. S.*, **19**, 237-238.

Ogata, Y. and Katsura, K. (1991) Maxmum likelihood estimates of the fractal dimension for random spatial patterns. *Biometrika*, **78**, 463-474.

Ogawa, T. (1989) Tessellations and fractals (Japanese). *Proc. Inst. Statist. Math.*, **37**, 107- 118.

Ohta, S. (1989) Pattern formation with anisotropy and quantitative analysis (Japanese). *Proc. Inst. Statist. Math.*, **37**, 81-88.

Okamoto, M. (1989) Comments on "Long-memory models and their statistical properties" (Japanese). *J. Japn. Statist. S.*, **19**, 239-240.

Onural, L. (1991) Generating connected textured fractal patterns using Markov random fields. *IEEE Trans. Pattern Anal. Mach. Int.*, **13**, 819-825.

Oodaira, H. (1987) Discussion of session on self-similar processes. *Proceedings of the 46th Session of ISI, Tokyo, Book 4*, pp. 256-257.

Oodaki, M. (1993) On the invertibility of fractionally differenced ARIMA processes. *Biometrika*, **80, No. 3**, 703-709.

Ortega, J. (1989) Upper classes for the increments of fractional Wiener processes. *Probab. Th. Rel. Fields*, **80**, 365-379.

Pages, G. (1987) Détection de changements brusques des charactéristiques spectrales d'un signal numérique. *Cahier du CERMA*, pp. 76-99.

Pakshirajan, R.P. and Vasudeva, R. (1981) A functional law of the iterated logarithm for a class of subordinators. *Ann. Probab.*, **9**, 1012-1018.

Parzen, E. (1974) Some recent advances in time series modelling. *IEEE Trans. Automatic Control*, **AC-19**, 723-729.

Parzen, E. (1981) Time series model identification and prediction variance horizon. *Applied Time Series Analysis II, D.F. Findley (ed.), Academic Press, New York*, 415-447.

Parzen, E. (1983) Time series model identification by estimating information, memory and quantiles. *Tech. Rep., Dept. of Statistics, Texas A&M University.*

Parzen, E. (1986) Quantile spectral analysis and long-memory time series. *J. Appl. Probab.*, **23A**, 41-54.

Parzen, E. (1992) Comment on "Statistical methods for data with long-range dependence." *Statist. Sci.*, **7, No. 4**, 420.

Passoja, D.E. and Amborski, D.J. (1978) Fractal profile analysis by Fourier transform methods. *Microstruct. Sci.*, **6**, 143-148.

Passoja, D.E. and Psioda, J.A. (1981) Fourier transform techniques - fracture and fatigue. *Fractography and Material Science, L.N. Gilbertson and R.D. Zipp (eds.), Am. Soc. for Testing Materials*, pp. 335-386.

Patzschke, N. and Zähle, M. (1992) Fractional differentiation in the self-affine case. I – Random functions. *Stoch. Proc. Appl.*, **43**, 165-175.

Pearson, E.F. and Wishart, J. (eds.) (1936) Student's collected papers. *Cambridge University Press, Cambridge.*

Pearson K. (1902) On the mathematical theory of errors of judgement, with special reference to the personal equation. *Philos. Trans. Roy. Soc. Series A*, **198**, 235-299.

Peirce, C.S. (1873) Theory of errors of observations. *Report of Superintendent of US Coast Survey (for the year ending November 1, 1870), Washington, D.C., Government Printing Office, Appendix No. 21, pp. 200-224 and Plate No. 27.*

Peiris, M.S. and Perera, B.J.C. (1988) On prediction with fractionally differenced ARIMA models. *J. Time Ser. Anal.*, **9**, 215-220.

Peitgen, H.-O. and Richter, P.H. (1986) The beauty of fractals. *Springer, New York.*

Peitgen, H.-O. and Saupe, D.(eds.) (1988) The science of fractal images. *Springer, New York.*

Peng, C.-K., Buldyrev, S.V., Goldberger, A.L., Havlin, S., Sciortino, F., Simons, M,. and Stanley, H.E. (1992) Long-range correlations in nucleotide sequences. *Nature*, **356**, 168-170.

Percival, D.B. (1977) Prediction error analysis of atomic frequency standards. *Proc. 31st Annual Symposium on Frequency Control, Atlantic City, NJ*, 319-326.

Percival, D.B. (1983) The statistics of long-memory processes. *PhD thesis, Dept. of Statistics, University of Washington, Seattle.*

Percival, D.B. (1985) On the sample mean and variance of a long memory process. *Tech. Rep. No. 69, Department of Statistics, University of Washington, Seattle.*

Pham, D.T. and Guégan, D. (1993) Asymptotic normality of the discrete Fourier transform of long memory time series. *preprint.*

Picard, D. (1985) Nonparametric statistical procedures for the change-point problem. *Adv. Appl. Probab.*, **17, No. 4**, 841-867.

Pollak, M., Croarkin, C., and Hagwood, C. (1993) Surveillance schemes with applications to mass calibration. *NIST Tech. Rep. 5158.*

Porter-Hudak S. (1990) An application of the seasonal fractionally differenced model to the monetary aggregates. *J. Am. Statist. Assoc.*, **85**, 338-344.

Portnoy, S.L. (1977) Robust estimation in dependent situations. *Ann. Statist.*, **5**, 22-43.

Priestley, M.B. (1981) Spectral analysis of time series. *Academic Press, London.*

Raftery, A.E. (1992) Comment on "Statistical methods for data with long-range dependence." *Statist. Sci.*, **7, No. 4**, 421-422.

Ramanathan, J. and O. Zeitouni, O. (1991) On the wavelet transform of fractional Brownian motion. *IEEE Trans. Inf. Th.*, **37**, 1156-1158.

Ramsey, F. (1974) Characterization of the partial autocorrelation function. *Ann. Statist.*, **2**, 1296-1303.

Ray, B.K. (1991) Forecasting long memory processes under misspecification. *ASA Proc. Bus. Econ. Statist. Sect.*, pp. 237-242.

Ray, B.K. (1993) Modeling long memory processes for optimal long-range prediction. *J. Time Ser. Anal.*, **14**, 511-526.

Reeve, R. (1992) A warning about standard errors when estimating the fractal dimension. *Comp. Geosci.*, **18**, 89-91.

Reisen, V.A. (1994) Estimation of the fractional difference parameter in the ARIMA(p, d, q) model using the smoothed periodogram. *J. Time Ser. Anal.*, **15, No. 3**, 335-350.

Renshaw, E. (1988) The high-order autocovariance structure of the telegraph wave. *J. Appl. Probab.*, **25**, 744-751.

Renshaw, E. (1994) The linear spatial-temporal interaction process and its relation to $1/\omega-$ noise. *J. Roy. Statist. Soc., Series B*, **56, No. 1**, 75-91.

Ripley, B. (1987) Stochastic simulation. *Wiley, New York.*

Rissanen, J. (1978) Modelling by shortest data description. *Automatica*, **14**, 465-471.

Rissanen, J. (1983) A universal prior for integers and estimation by minimum description length. *Ann. Statist.*, **11**, 416-431.

Robert, A. (1991) Fractal properties of simulated bed profiles in coarse-grained channels. *Math. Geol.*, **23**, 367-382.

Robinson, P.M. (1978) Alternative models for stationary stochastic processes. *Stoch. Proc. Appl.*, **8**, 141-152.

Robinson, P.M. (1989) Hypothesis testing in semiparametric and nonparametric models for economic time series. *Rev. Econ. Stud.*, **56**, 511-534.

Robinson, P.M. (1991a) Nonparametric function estimation for long-memory time series. *Nonparametric and semiparametric methods in econometrics and statistics, W. Barnett, J. Powell and G. Tauchen (eds.), Cambridge University Press, New York*, pp. 437-457.

Robinson, P.M. (1991b) Automatic frequency-domain inference on semi-parametric and nonparametric models. *Econometrica*, **59**, 1329-1364.

Robinson, P.M. (1991c) Testing for strong serial correlation and dynamic conditional heteroskedasticity in multiple regression. *J. Econometrics*, **47**, 67-84.

Robinson, P.M. (1991d) Semiparametric analysis of long-memory time series. *preprint.*

Robinson, P.M. (1991e) Consistent semiparametric estimation of long-range dependence. *preprint.*

Robinson, P.M. (1991f) Rates of convergence and optimal bandwidth in spectral analysis of processes with long-range dependence. *preprint.*

Robinson, P.M. (1992a) Log-periodogram regression for time series with long-range dependence. *preprint.*

Robinson, P.M. (1992b) Optimal tests of nonstationary hypothesis. *preprint.*

Robinson, P.M. (1993a) Highly insignificant F-ratios. *preprint.*

Robinson, P.M. (1993b) Time series with strong dependence. *Adv. in Econometrics, 76, 6th World Congress of the Econometric Society, Vol. 1, Cambridge University Press, Cambridge.*

Robinson, P.M., Bera, A.K., and Jarque, C.M. (1985) Tests for serial dependence in limited dependent variable models. *Int. Econ. Rev.*, **26**, 629-638.

Rocke, D.M. (1983) Robust statistical analysis of interlaboratory studies. *Biometrika*, **70**, 421-431.

Rohatgi, V.K. (1982) Operator self-similarity. *Nonpar. Statist. Inf., B.V. gnedenko, M.L. Puri and I. Vincze (eds.), Elsevier/N. Holland*, pp. 773-778.

Rosen, J. (1987) The intersection local time of fractional Brownian motion in the plane. *J. Multivar. Anal.* , **23**, 37-46.

Rosenblatt, M. (1961) Independence and dependence. *Proc. 4th Berkeley Symp., Univ. California, Berkeley Univ. Press*, pp. 431-443.

Rosenblatt, M. (1976) Fractional integrals of stochastic processes and the central limit theorem. *J. Appl. Probab.*, **13**, 723-732.

Rosenblatt, M. (1979) Some limit theorems for partial sums of quadratic forms in stationary Gaussian variables. *Z. Wahrsch. verw. Geb.*, **49**, 125-132.

Rosenblatt, M. (1981) Limit theorems for Fourier transforms of functionals of Gaussian sequences. *Z. Wahrsch. verw. Geb.*, **55**, 123-132.

Rosenblatt, M. (1984) Stochastic processes with short-range and long-range dependence. *Statistics: An Appraisal. Proceedings 50th Anniversary Conference. H.A. David, H.T. David (eds.). The Iowa State University Press*, pp. 509-520.

Rosenblatt, M. (1985) Stationary sequences and random fields. *Birkhäuser, Basel.*

Rosenblatt, M. (1987) Scale renormalization and random solutions of the Burgers equation. *J. Appl. Probab.*, **24**, 332-338.

Rosenblatt, M. (1991) Stochastic curve estimation. *Inst. Math. Statist., California.*

Rosinsky, J. (1986) On stochastic integral representation of stable processes with sample paths in Banach spaces. *J. Multivar. Anal.*, **20**, 277-302.

Rousseeuw, P.J. and Leroy, A.M. (1987) Robust regression and outlier detection. *Wiley, New York.*

Rozanov, Yu. A. (1967) Stationary random processes. *Holden-Day, San Francisco.*

Samarov, A. and Taqqu, M.S. (1988) On the efficiency of the sample mean in long-memory noise. *J. Time Series Anal.*, **9**, 191-200.

Samorodnitsky, G. (1990) Integrability of stable processes. *Tech. Report No. 301, Dept. of Statistics, Univ. of North Carolina, Chapel Hill.*

Samorodnitsky, G. and Taqqu, M.S. (1989) The various fractional Lévy motions. *Probability, Statistics and Mathematics, Papers in Honour of Samuel Karlin. T.W. Anderson, K.B. Athreya, D.L. Inglehart (eds.). Academic Press, Boston*, pp. 261-270.

Samorodnitsky, G. and Taqqu, M.S. (1990) $1/\alpha$−self-similar processes with stationary increments. *J. Multiv. Anal.*, **35**, 308-313.

Samorodnitsky, G. and Taqqu, M.S. (1992) Linear models with long-range dependence and finite or infinite variance. *New Directions in Time Series Analysis, Part II, D. Brillinger, P. Caines, J. Geweke, E. Parzen, M. Rosenblatt and M.S. Taqqu (eds.), IMA Volumes in Mathematics and its Applications, Vol. 46, Springer, New York*, pp. 325-340.

Samorodnitsky, G. and Taqqu, M.S. (1993) Stable random processes. *In press.*

Sargan, J.D. and Bhargawa, A. (1983) Testing residuals from least squares regression for being generated by the Gaussian random walk. *Econometrica*, **51**, 153-174.

Sarpkaya, T. and Isaacson, M. (1981) Mechanics of wave forces on offshore structures. *Van Nostrand Reinhold, New York.*

Sato, K. (1991) Self-similar processes with independent increments. *Probab. Th. Rel. Fields*, **89**, 285-300.

Sato, Y. (1989) Joint distributions of some self-similar stable processes. *preprint.*

Sato, Y. and Takenaka, S. (1991) On determinism of symmetric α−stable processes of generalized Chentsov type. *preprint.*

Sayles, R.S. and Thomas, T.R. (1978) Surface topography as a nonstationary random process. *Nature*, **271**, 431-434.

Scheffé H. (1959) The analysis of variance. *Wiley, New York*

Scheuring. I. (1991) The fractal nature of vegetation and the species-area relation. *Theor. Pop. Biol.*, **39**, 170-177.

Schick, K.L. and Verveen, A.A. (1974) $1/f$ noise with a low frequency white noise limit. *Nature*, **251**, 599-601.

Schmidt, C, and Tschernig, R. (1993) Identification of fractional ARIMA models in the presence of long memory. *Münchner Wirtschaftswissenschaftliche Beiträge, Nr. 93-04, Luidwig-Maximilians-Universität, Munich.*

Schroeder, M. (1991) Fractals, chaos, power laws. *Freeman, New York.*

Schulman, E. (1956) Dendoclimatic changes in semiarid America. *University of Arizona Press, Tucson.*

Schwarz, G. (1978) Estimating the dimension of a model. *Ann. Statist.*, **6**, 461-464.

Seiler, F.A. (1986) Use of fractals to estimate environmental dilution factors in river basins. *Risk Anal.*, **6**, 15-26.

Serfling, R. (1980) Approximation theorems of mathematical statistics. *Wiley, New York.*

Seshadri, V. and West, B.J. (1982) Fractal dimensionality of Levy processes. *Proc. Natl. Acad. Sci. USA*, **79**, 4501-4505.

Sethuraman, S. and Basawa, I.V. (1992) Parameter estimation in a stationary autoregressive process. *Topics in Stochastic Inference, I.V. Basawa and N.U. Prabhu (eds.), In press.*

Shea, G.S. (1989) Regression estimation and bootstrap inference on the order of fractional integration in multivariate time series models. *preprint.*

Shea, G.S. (1991) Uncertainty and implied variance bounds in long-memory models of the interest rate term structure. *Empirical Economics*, **16**, 287-312.

Shibata R. (1976) Selection of the order of an autoregressive model by Akaike's information criterion. *Biometrika*, **63**, 117-126.

Shibata R. (1980) Asymptotically efficient selection of the order of the model for estimating parameters of a linear process. *Ann. Statist.*, **8**, 147-164.

Siddiqui, M.M. (1976) The asymptotic distribution of the range and other functions of partial sums of stationary processes. *Water Resources Res.*, **12**, 1271-1276.

Simon, B. (1974) The $P(\phi)_2$ Euclidian (Quantum) field theory. *Princeton University Press, Princeton.*

Sims, C.A. (1988) Bayesian skepticism on unit-root econometrics. *J. Econ. Dyn. Control*, **12**, 463-474.

Sinai Ya.G. (1976) Self-similar probability distributions. *Theory Probab. Appl.*, **21**, 64-80.

Smalley, Jr, R.F., Chatelain, J.-L., Turcotte, D.L., and Prevot, R. (1987) A fractal approach to the clustering of earthquakes: Applications to the seismicity of the new hebrides. *Bull. Seism.*, **77**, 1368-1381.

Smith, H.F. (1938) An empirical law describing heterogeneity in the yields of agricultural crops. *J. Agric. Sci.*, **28**, 1-23.

Smith, R.L. (1991) Estimating dimensions in noisy chaotic time series. *J.Roy. Statist. Soc., Series B*, **54**, 329-351.

Smith, R.L. (1992a) Optimal estimation of fractal dimension. *Nonlinear Modeling and Forecasting, M. Casdagli and S. Eubank (eds.), Addison Wesley*, pp. 115-135.

Smith, R.L. (1992b) Comment on "Statistical methods for data with long-range dependence." *Statist. Sci.*, **7, No. 4**, 422-425.

Smith, R.L. (1992c) Comment on "Chaos, fractals and statistics." *Statist. Sci.*, **7**, 109-113.

Smith, R.L. (1993) Long-range dependence and global warming. *Statistics for the Environment, V. Barnett and F. Turkman (eds.), Wiley, Chichester.*

Solo, V. (1989a) Intrinsic random functions and the paradox of $1/f$ noise. *preprint.*

Solo, V. (1989b) Asymptotics of quadratic forms for time series. *preprint.*

Sowell, F.B. (1986) Fractionally integrated vector time series. *PhD thesis, Duke University.*

Sowell, F.B. (1990) The fractional unit-root distribution. *Econometrica*, *58*, 495-505.

Sowell, F. (1992a) Modelling long run behavior with the fractional ARIMA model. *J. Monet. Econ.*, **29**, 277-302.

Sowell, F.B. (1992b) Maximum likelihood estimation of stationary univariate fractionally integrated time series models. *J. Econometrics*, **53**, 165- 188.

Spitzer, F. (1969) Uniform motion with elastic collision of an infinite particle system. *J. Math. Mech.*, **18**, 973-989.

Spolia, S.K., Chander, S., and O'Connor, K.M. (1980) An autocorrelation approach for parameter estimation of fractional order equal-root autoregressive models using hypergeometric functions. *J. Hydrol.*, **47**, 1-18.

Steele, J.M. Shepp, L.A., and Eddy, W.F. (1987) On the number of leaves of a Euclidian minimal spanning tree. *J. Appl. Probab.*, **24**, 809-826.

Steiger, W. (1964) A test of nonrandomness in stock price changes. *The random character of stock market prices, P.H. Cootner (ed.), MIT Press, Cambridge, Mass.*

Stone, C. (1963) Zeros of semi-stable processes. *Illinois J. Math.*, **7**, 631-637.

Stuck, B.W. and Kleiner, B. (1974) A statistical analysis of telephone noise. *The Bell System Technical Journal*, **53**, 1263-1320.

Student (1927) Errors of routine analysis. *Biometrika*, **19**, 151-164.

Sullivan, F. and Hunt, F. (1988) How to estimate capacity dimension. *Nuclear Phys. B (Proc. Suppl.)*, **5A**, 125-128.

Sun, T.C. (1963) A central limit theorem for non-linear functions of a normal stationary process. *J. Math. Mech.*, *12*, 945-978.

Sun, T.C. (1965) Some further results on central limit theorems for nonlinear functions of normal stationary processes. *J. Math. Mech.*, **14**, 71-85.

Sun, T.C. and Ho, H.C. (1986) On central and non-central limit theorems for non-linear functions of a stationary Gaussian process. *Dependence in Probability and Statistics, E. Eberlein and M.S. Taqqu (eds.). Birkhäuser, Boston*, pp. 3-19.

Surgailis, D. (1981a) Convergence of sums of nonlinear functions of moving averages to self-similar processes. *Soviet Math.*, **23**, 247-250.

Surgailis, D. (1981b) Zones of attraction of self-similar multiple integrals. *Lithuanian Math. J.*, **22**, 327-340.

Surgailis, D. (1981c) On L^2 and non-L^2 multiple stochastic integration. *Stochastic Differential Systems, Lecture Notes Control Inf. Sci., Vol. 36, Springer, New York*.

Surgailis, D. (1981d) On infinitely divisible self-similar random fields. *Z. Wahrsch. verw. Geb.*, **58**, 453-477.

Takeda, S. (1960) An investigation of airplane loads during taxiing. *Monthly Reports of Transportation Technical Research Institute (in Japanese)*, **10, No. 5**.

Takenaka, S. (1991a) Integral-geometric construction of self-similar stable processes. *Nagoya Math. J.*, **123**, 1-12.

Takenaka, S. (1991b) Examples of self-similar stable processes. *preprint*.

Taqqu, M.S. (1970) Note on the evaluation of R/S for fractional noises and geophysical records. *Water Resources Res.*, **6**, 349-350.

Taqqu, M.S. (1972) Limit theorems for sums of strongly dependent random variables. *PhD thesis, Columbia University*.

Taqqu, M. S. (1975) Weak convergence to fractional Brownian to the Rosenblatt process. *Z. Wahrsch. verw. Geb.*, **31**, 287-302.

Taqqu, M.S. (1977) Law of the iterated logarithm for sums of non-linear functions of Gaussian random variables. *Z. Wahrsch. verw. Geb.*, **40**, 203-238.

Taqqu, M.S. (1978a) A representation for self-similar processes. *Stoch. Proc. Appl.*, **7**, 55-64.

Taqqu, M.S. (1978b) Weak convergence at all Hermite ranks. *Tech. Rep. No. 389, School of Operations Research, Cornell University*.

Taqqu, M.S. (1979) Convergence of integrated processes of arbitrary Hermite rank. *Z. Wahrsch. verw. Geb.*, **50**, 53-83.

Taqqu, M.S. (1981) Self-similar processes and related ultraviolet and infrared catastrophes. *Random Fields: Rigorous Results in Statistical Mechanics and Quantum Field Theory. Colloquia Mathematics Soci-*

etatis Janos Bolai, Vol. 27, Book 2, North Holland, Amsterdam, pp. 1057-1096.

Taqqu, M.S (1986a) Sojourns in an elliptical domain. *Stoch. Proc. Appl.*, **21**, 319-326.

Taqqu, M.S. (1986b) A bibliographical guide to self-similar processes and long-range dependence *Dependence in probability and statistics. E. Eberlein and M.S. Taqqu (eds.), Birkhüser, Boston*, pp. 137-165.

Taqqu, M.S. (1987) Random processes with long-range dependence and high variability. *J. Geophys. Res.*, **92**, 9682-9686.

Taqqu, M.S. (1988) Self-similar processes. *Encyclopedia of Statistical Sciences, S. Kotz and N. Johnson, Vol. 8, Wiley, New York*, 352-357.

Taqqu, M.S. and Czado, C. (1985) A survey of functional laws of the iterated logarithm for self-similar processes. *Stoch. Models*, **1**, 77-115.

Taqqu, M.S. and Levy, J. (1986) Using renewal processes to generate long-range dependence and high variability. *Dependence in Probability and Statistics, E. Eberlein and M.S. Taqqu (eds.), Birkhäuser, Boston*, pp. 73-89.

Taqqu, M.S. and Wolpert, R. (1983) Infinite variance self-similar processes subordinate to a Poisson measure. *Z. Wahrsch. Verw. Geb.*, **62, No. 1**, 53-72.

Taylor, C.C. and Taylor, S.J. (1991) Estimating the dimension of a fractal. *J. Royal Statist. Soc., Series B*, **53**, 353-364.

Taylor, G.I. (1935) Statistical theory of turbulence. *Proc. Roy. Soc. Ser. A*, **151**, 421-478.

Taylor, S.J. (1973) Sample path properties of processes with stationary independent increments. *Stochastic Analysism D.G. Kendall and E.F. Harding (eds.), Wiley, Ne York*, pp. 387-414.

Telcs, A. (1989) Random walks on graphs, electric networks and fractals. *Probab. Th. Rel. Fields*, **82**, 435-449.

Telcs, A. (1990) Spectra of graphs and fractal dimensions. *Probab. Th. Rel. Fields*, **85**, 489-497.

Teman, R. (ed.) (1976) Turbulence and Navier Stokes equations. *Springer Lecture Notes in Math., Vol. 565, Springer, New York*.

Terrin, N. and Hurvich, C. (1992) An asymptotic Wiener-Ito representation for the low frequency ordinates of the periodogram of a long-memory time series. *preprint.*

Terrin, N. and Taqqu, M.S. (1990a) A noncentral limit theorem for quadratic forms of Gaussian stationary sequences. *J. Theor. Probab.*, **3**, 449-475.

Terrin, N. and Taqqu, M.S. (1990b) Quadratic forms with long-range dependence. *Proc. Vilnus, Vol. 2*, pp. 466-473.

Terrin, N. and Taqqu, M.S. (1991a) Power counting theorem in Euclidean space. *Random Walks, Brownian Motion and Interacting Particle Systems. Festschrift/Frank Spitzer. R. Durrett and H. Kesten (eds.), Birkhäuser, Basel*, pp. 425-440.

Terrin, N. and Taqqu, M.S. (1991b) Convergence in distribution of sums of bivariate Appell polynomials with long-range dependence. *Probab. Th. Rel. Fields*, **90**, 57-81.

Terrin, N. and Taqqu, M.S. (1991c) Convergence to a Gaussian limit as the normalization exponent tends to 1/2. *Statist. Probab. Lett.*, **11**, 419-427.

Tewfik, A.H. and Kim, M. (1992) Correlation structure of the discrete wavelet coefficients of fractional Brownian motion. *IEEE Trans. Inf. Th.*, **38**, 904-909.

Thomas, T.R. and Thomas, A.P. (1988) Fractals and engineering surface roughness. *Surf. Topogr.*, **1**, 143-152.

Thompson, W.E. (1958) Measurements and power spectra of runway roughness at airports in countries of the North Atlantic Treaty Organization. *NACA TN 4303*.

Tong, H. (1993) Non-linear time series. *Oxford University Press*.

Tousson, O. (1925) Mémoire sur l'Histoire du Nil. *Mémoires de l'Institut d'Egypte*.

Toyoki, H. (1989) Self-similarity of ordering process in a system containing point-type topological defects (Japanese). *Proc. Inst. Statist. Math.*, **37**, 89- 98.

Troutman, B.M. (1983) Weak convergence of the adjusted range of cumulative sums of exchangeable random variables. *J. Appl. Probab.*, **20**, 297- 304.

Tsay, R.S. (1992) Comment on "Chaos, fractals and statistics." *Statist. Sci.*, **7**, 113-114.

Tschernig, R. (1992) Wechselkurse, Unischerheit und Long-Memory. *PhD dissertation, University of Munich*.

Tschernig, R. (1993a) Long memory in foreign exchange rates revisited. *preprint*.

Tschernig, R. (1993b) Detecting small sample bias in ARFIMA models. *preprint*.

Tschernig, R. and Zimmermann, K.F. (1992) Illusive persistence in German unemployment. *Recherches Economiques de Louvain*, **3-4**, 441-453.

Turcotte, D.L. (1986) Fractals and fragmentation. *J. Geophys. Res.*, **91**, 1921-1926.

Usami, Y. and Nagatani. T. (1989) Aggregation in coupled diffusion fields: A model for electrochemical deposition (Japanese). *Proc. Inst. Statist. Math.*, **37**, 189-197.

van Harn, K. and Steutel, F.W. (1985) Integer-valued self-similar processes. *Comm. Statist. Stoch. Models*, **1**, 191-208.

Vervaat, W. (1985) Sample path properties of self-similar processes with stationary increments. *Ann. Probab.*, **13**, 1-27.

Vervaat, W. (1986) Stationary self-similar extremal processes and random semicontinuous functions. *Dependence in Probability and Statis-*

tics, E. Eberlein and M.S. Taqqu (eds.), *Birkhäuser, Boston*, pp. 457-470.

Vervaat W. (1987) Properties of general self-similar processes. *Bull. Int. Statist. Inst.*, **52, No. 4**, 199-216.

Vitale, R.A. (1973) An asymptotically efficient estimate in time series analysis. *Quat. Appl. Math.*, **30**, 421-440.

Voldman, J., B. Mandelbrot, B.B., Hoevel, L.W., J. Knight, J. and Rosenfeld, P. (1983) Fractal nature of software-cache interaction. *IBM J. Res. Dev.*, **27**, 164-170.

Voss, R.V. (1979) $1/f$ flicker noise: A brief review. *Proc. 33rd Ann. Symp. on Frequency Control, Atlantic City*, pp. 40-46.

Voss, R.F. (1979) $1/f$ (flicker) noise: A brief review. *Proc. 33rd Annual Symp. on Frequency Control, Atlantic City, NJ*, pp. 40-46.

Voss, R.F. and Clarke, J. (1975) $1/f$ noise in music and speech. *Nature*, **258**, 317-318.

Vuolle-Apiala, J. (1989) Time-changes of self-similar Markov processes. *Ann. Inst. H. Poincaré Probab. Statist.*, **25**, 581-587.

Vuolle-Apiala, J. and Graversen, S.E. (1986) Duality theory for self-similar processes. *Ann. Inst. H. Poincaré Probab. Statist.*, **22**, 323-332.

Wallis, J.R. and Matalas, N.C. (1970) Small sample properties of H and K, estimators of the Hurst coefficient h. *Water Resources Res.*, **6**, 1583-1594.

Wallis, J.R. and Matalas, N.C. (1971) Correlogram analysis revisited. *Water Resources Res.*, **7**, 1448-1459.

Wallis, J.R. and Matalas, N.C. (1972) Correlogram analysis revisited. *Water Resources Res.*, **8**, 1112-1117.

Wallis, J.R. and O'Connell, P.E. (1973) Firm reservoir yield - how reliable are historical records ? *Hydrol. Sci. Bull.*, **XVIII**, 347-365.

Walls, J.H., Houbolt, J.V., and Press, H. (1954) Some measurements and power spectra of runway roughness. *NACA TN 3305*.

Watson, G.S. (1967) Linear least squares regression. *Ann. Math. Statist.*, **28**, 1679-1699.

Waymire, E. (1985) Scaling limits and self-similarity in precipitation fields. *Water Resour. Res.*, **21**, 1271-1281.

Waymire, E., Gupta, K. and Rodriguez-Iturbe, I. (1984) A spectral theory of rainfall intensity at the mese-β scale. *Water Resour. Res.*, **20**, 1453-1465.

Weitz, D.A., Sander, L.M., and Mandelbrot, B.B. (eds.). Fractal aspects of materials. *Material Res. Soc., Pittsburgh*.

Welch, P.D. (1967) The use of the fast Fourier transform for estimation of spectra: A method based on time averaging over short, modified periodograms. *IEEE Trans. Electr. Acoust. AU*, **15**, 70.

Wendel, J.G. (1984) Zero-free intervals of semi-stable Markov processes. *Math. Scand.*, **14**, 21-34.

Weron, A. (1984) Stable processes and measures, a survey. *Prob. Th. in Vector Spacess III, Lecture Notes in Math., 1080, Springer, New York*, pp. 306-364.

Whalley, W.B. and Orford, J.D. (1989) The use of fractals and pseudofractals in the analysis of two-dimensional outlines: Review and further exploration. *Cmp&Geos*, **15**, 185-197.

Whistler, D.E.N. (1990) Semiparametric models of daily and intra-daily exchange rate volatility. *PhD thesis, University of London.*

Whittle, P. (1951) Hypothesis testing in time series analysis. *Hafner, New York.*

Whittle, P. (1953) Estimation and information in stationary time series. *Ark. Mat.*, **2**, 423-434.

Whittle, P. (1956) On the variation of yield variance with plot size. *Biometrika*, **43**, 337-343.

Whittle, P. (1962) Topographic correlation, power-law covariance functions, and diffusion. *Biometrika*, **49**, 304-314.

Willett, H.C. (1964) Evidence of solar climatic relationships. *Weather and Our Food Supply, Iowa State University, Ames*, pp. 123-151.

Williams, R.M. (1952) Experimental design for serially correlated observations. *Biometrika*, **39**, 151-167.

Wilson, E.B. and Hilferty, M.M. (1929) Note on C.S. Peirce's experimental discussion of the law of errors. *Proc. Natl. Acad. Sci. USA*, **15, No. 2**, 120-125.

Wiseman, N.E. and Nedunuri, S. (1986) Computing random fields. *Comput. J.*, **29**, 373-377.

Wolf, D. (ed.) (1978) Noise in physical systems. *Proc. 5th Int. Conf. on Noise, Bad Nauheim, Springer Series in Electro-Physics, Vol. 2, Springer, New York.*

Working, H. (1934) A random difference series for use in the analysis of time series. *J. Am. Statist. Assoc.*, **29**, 11-14.

Xian-Yin, Z. (1992) On the recurrence of simple random walks on some fractals. *J. Appl. Probab.*, **29**, 454-459.

Yajima, Y. (1985) On estimation of long-memory time series models. *Austral. J. Statist.*, **27**, 303-320.

Yajima, Y. (1988) On estimation of a regression model with long-memory stationary errors. *Ann. Statist.*, **16**, 791-807.

Yajima, Y. (1989a) A central limit theorem of Fourier transforms of strongly dependent stationary processes. *J. Time Ser. Anal.*, **10**, 375-383.

Yajima, Y. (1989b) Long-memory models and their statistical properties (Japanese). *J. Japn. Statist. S.*, **19**, 219-236.

Yajima, Y. (1989c) Reply to comments on "Long-memory models and their statistical properties" (Japanese). *J. Japn. Statist. S.*, **19**, 244-246.

Yajima, Y. (1991) Asymptotic properties of the LSE in a regression model with long-memory stationary errors. *Ann. Statist.*, **19**, 158-177.

Yakubo, K. and Nakayama, T. (1989) Dynamics of fractals (Japanese). *Proc. Inst. Statist. Math.*, **37**, 47-61.

Yates, F. (1949, 1981) Sampling methods for censuses and surveys. *Griffin, London.*

Yates, F. and Finney, D.J. (1942) Statistical problems in field sampling for wireworms. *Ann. Appl. Biol.*, **29**, 156-167.

Youden, W.J. (1972) Enduring values. *Technometrics*, **14**, 1-11.

Young, W.E. (1971) Random walk of stock prices; a test of the variance-time function. *Econometrica*, **39**, 797-812.

Zähle, U. (1984) Random fractals generated by random cutouts. *Mat. Nachr.*, **116**, 27- 52.

Zähle, U. (1988a) Self-similar random measures I. Notion, carrying Hausdorff dimension, and hyberbolic distribution. *Probab. Th. Rel. Fields*, **80**, 79-100.

Zähle, U. (1988b) The fractal character of localizable measure-valued processes II. Localizable processes and backward trees. *Mat. Nachr.*, **137**, 35-48.

Zähle, U. (1990) Self-similar random measures II. A generalization: Self-affine measures. *Mat. Nachr.*, **146**, 85-98.

Zamar, R. (1990) Robustness against unexpected dependence in the location model. *Statist. Probab. Lett.*, **9**, 367-374.

Zawadzki, I.I. (1973) Statistical properties of precipitation patterns. *J. Appl. Meteorol.*, **12**, 459-472.

Zygmund, A. (1953) Trigonometric series. *Cambridge University Press, London.*

Author index

Subject index

DATE DUE